Make Your Own Global Warming

Using HAARP, Chemtrails, and the Sun

By Steve Preston

1st Edition

Table of Contents

Introduction

I know you are hearing all types of conspiracy theories that are making you confused about our safety. You try to ignore them, but people whisper them in your ears so I thought the first thing to do would be to hit a few right away and reduce their credibility as we go along so you can begin to have a better appreciation of global stability and how we can *"Save our Planet"*. I love those little slogans just like you so I may say a few as we go along. We will also look at how to change the weather patterns of the earth outside these conspiracies.

- **Ozone Destruction Conspiracy**-Here is one conspiracy I think should be halted right now. Some dodo head claimed that CO2 being poured into the atmosphere affected the amount of Ozone that protects us from UV-B rays from the sun so that everyone would get cancer because we have too much carbon dioxide. These people didn't even look up to see what OZONE was and still pushed the limits of absurdity? Ozone is made when one of the oxygen atoms of CO_2 is knocked off and it attaches to regular oxygen. Without CO_2----no Ozone.
- **The Coal Conspiracy**- Another knuckle head decided that one of the most benign energy

absorbing molecules around would absorb the sun's energy and heat up the planet. Here is the kicker. They called it a green gas [A reference to having something to do with plants, but plants give off Oxygen and produce sugars not carbon dioxide. They simply use water, nitrates, CO_2 and sunlight]. You would think the first thing to do would be to change it to "anti-green gas" or, "people gas" or even "all gas", but here is the real kicker. Here comes one of those slogans *"GO GREEN-I mean remove what makes things green".* Just about everyone knows only 0.02% of our air is carbon dioxide and it simply cannot absorb solar energy for heating AT ALL. The conspiracists cleverly convinced many somewhat shortsighted "scientist want-to-bes" and continued by saying the Earth regenerates energy in the form of longer wave heat which is absorbed by CO_2 to make us even hotter. I don't know what type of physics they use but the retransmitted energy from the earth is mostly absorbed by 2 things Oxygen and water vapor which make up most of the non-nitrogen or Argon molecules in the air. The whole thing was predicated on a number of lies as sort of a cover up, but hopefully, I will be able to enlighten the reader on the absurdity and reasons we are seeing temporary and sporadic heating in the Arctic that these knuckleheads pull out to confound many.

- **Manmade Global Warming Conspiracy-** I know you have heard this one. Everything from letting

Freon escape from your refrigerator, using underarm spray, and riding in a fuel guzzling jet were destroying our cool landscape. A Senator is now trying to put people in jail that won't stop using their automobiles or heating with nasty coal. The ones he really wants in jail are the oil company officials as he has this conspiracy in his head that oil company magicians somehow make those yelling about mankind burning up the earth with jet-fuel look like idiots using data that should not be allowed to be used. You'll love the absurdity of this as almost 100 disaster models show how driving a car would make the oceans flood the land and temperatures rage out of control by now. This one was especially sinister as they had to get the quasi-scientific community to attest to data that was KNOWN to be false. Even President Obama got in on this one as he shut done over a hundred Coal plants saying he did it for the GOOD of America.

- **Venusian Greenhouse Catastrophe Conspiracy-** You won't believe this one unless you heard it before and had been drinking. One group called NOAA said because Venus has CO_2 on it, whatever made it get hot must have been runaway temperature caused by that CO_2, again calling it a greenhouse gas. Then they added the kicker--- if someone makes electricity with coal our planet will burn up just like Venus. They continue with even more craziness. *"Carbon dioxide is increasing so fast we will be in a <u>thermal runaway</u> condition in a few years unless we*

8

stop using coal and gas immediately." Can you imagine the criminal audacity and ignorance? They even brought out some aerosol CO_2 data collected from Hawaii to assure everyone CO_2 levels in the Pacific Ocean have gone from 0.02% to 0.025% over about 10 years and shuttered with fear as they told us the news. With only 3.2% of the Airborne CO_2 as the non-naturally occurring Carbon Dioxide, this means if we double our coal produced electricity use, we are talking about an increase of about 0.003%.

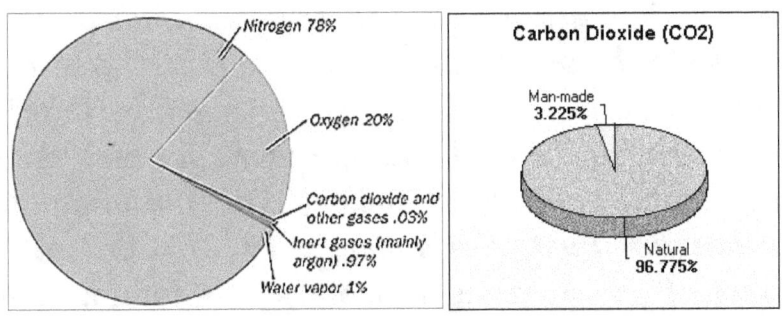

That didn't work well so they falsified Ice Core data to make it look like our earth was reaching a new level of instability by adding Hawaii data to Antarctican Ice data. The new level of subterfuge is really getting crazy with well-meaning people who are not shown all the data are so afraid they want those not trying to shut down the Coal Industry to be put in jail.

Global Warming–Washington Post Arctic Report- *The Arctic Ocean is warming up, icebergs are growing scarcer and in some places the seals are finding the water too hot, according to a report to the Commerce department yesterday from Consulate, at Bergen, Norway. Reports from fishermen, seal hunters and* *explorers all point to a radical change in climate conditions and hitherto unheard-of temperatures in the Arctic zone. Exploration expeditions report that scarcely any ice has been met as far north as 81 degrees 29 minutes. Soundings to a depth of 3,100 meters showed the Gulf Stream still very warm. Great masses of ice have been replaced by moraines of earth and stones, the report continued, while at many points well known glaciers have entirely disappeared. Very few seals and no white fish are found in the eastern Arctic, while vast shoals of herring and smelts which have never before ventured so far north, are being encountered in the old seal fishing grounds. **Within a few years it is predicted that due to the ice melt the sea will rise and make most coastal cities uninhabitable**. [Washington Post, November 2, 1922].* Certainly the number of gas guzzling Model-Ts was forcing almost extinction of the Arctic Ice and men were getting rich on solar cells.

While I would like to simply say don't worry about all of these conspiracies, we need to address them with a level of caution. After all, there is always a little bit of fact in a conspiracy. That is why so many fall for them so easily. What we need to do is trace some of them to an origin. For part of this conspiracy relief, we will need to address a couple of thing that have been clouded by conspiracy themselves these are the mysterious smoke trails behind some unusual Jetliners and the published work of the various Ionospheric Heater facilities around the world, especially the

HAARP one in Alaska and the EISCAT one in Norway as they are insanely powerful when it comes to being able to transmit dangerous ELF waves into our atmosphere, the Ionosphere, and even the delicate Magnetosphere.

I know some have simply gotten caught up in the "fever" of global warming and others have solace in empire building, but whenever you put on the name "scientist" you have a higher responsibility that is being trampled by the drama and lying. Before I get going let me say that there are signs of increased temperatures in the Arctic. What I get so mad about is some have demonized carbon dioxide as the culprit that is shifting our weather patterns. As we go along I will show you how that is impossible but there is another more logical villain.

Solar Villain

One "villain" we will investigate is the sun. Never think of the sun as some constant light ball of gas as there is no question that its cycles are being sensed on our planet in a big way. The other thing that must be viewed seriously is anything messing with our ionosphere and magnetosphere. That is just dangerous; especially in the less stable winter months. I know you are thinking this is some conspiracy book and some of it has that flavor, but in general, the detail presented here are simply de-falsified data and combination of evidence that tells us some shocking things. The book presents the issues, the data, the reasons and potential characterization of the outcomes given a number of possibilities. One thing that we have to investigate here is something called a Corona Mass Ejection from our sun. We have simply had too many close calls to ignore this even if it seems that everyone is focused on how to destroy the coal industry.

Contrails Versus Chem-trails and HAARP

I know you have heard of Contrails also known as chem-trails or vapor trails. Let me just say I am not talking about simple contrails which are just water vapor condensation trails behind the super-heated exhaust of a jet engine. We will be looking at trails that look similar, but instead of being simple vapor, they are filled with effluent on purpose. We will examine where these special "chem-trails", are found in the heaviest concentrations, read testimony from some who have firsthand knowledge and see how the chem-trails are now being used in conjunction with something called HAARP [High Frequency Active Auroral Research Program] emissions to bring about something that not only appears to be global warming, it is. I know this sounds bizarre and I have always stayed away from just reviewing work from people that appeared to look for conspiracy around every corner.

Something Sounded Fishy

Quite frankly I thought this chem-trail quagmire was one of the areas to stay away from. Chem-trails have been described since the 1940s and blamed on just about everything including deforestation, disease, hiding some sinister threat, reducing food, used as a weather weapon, reducing virility, and dirtying our water. Apparently, the effluents are simple hygroscopic metallic oxides like AlO_2, BaO. The HAARP thing has its own set of conspiracies including production of floods, chronic fatigue syndrome, power outages, all types of weather control anomalies around the world, the downing of flight 800, and worst of all; it was said to have destroyed the space shuttle Columbia in 2003. The conspiracy proponents not only included the normal nut cases, but also senators, congressmen, the Governor of Minnesota, the European Parliament, physicists, university professors, and investigative journalists. It seems everyone wanted to describe how some sinister agency or group of agencies was manipulating something with these 2 seemingly separate anomalous elements. How could this be; and the "why" was always problematic, but the search went on. This book will show some of the characteristics, oddness of occurrence, and strange outcomes surrounding events using both of these elements together, seemingly being controlled by our government and other superpower government. I don't know about all that. All I know is something fishy is going on

Global Warming

That gets me to global warming. We quit using underarm spray, would not recharge air conditioners, and converted to partial plant fuels for our automobiles all in an effort to thwart what we were told would be the end of the world. All the while, billions of dollars were

13

taken away from infrastructure of governments and converted to cash to line the pockets of just about anyone who claimed they could eliminate gasoline and coal around the world. After all, they pointed to a little molecule of carbon dioxide as the instigator of death to the world. Sometimes the temperature looked higher and then it would drop only to go higher again and allow for more funding. Sometimes ice melted and sometimes it came back with a great fanfare. Sometimes the temperature rose while CO_2 receded and when the CO_2 levels shot up, the temperature did not follow. That by itself could fuel a conspiracy, but I want to step back and look at characteristics of global warming, dense chem-trail effluents, and the activation of HAARP transmitters just to see where it leads. As a spoiler, let me just say it does not end well for us whether we place on it the title of conspiracy or just horrible facts.

Government Push

Even with 50% of our electrical energy coming from coal, we are being pushed into eliminating the use of burning coal and even with horrible inefficiencies and increase wear of auto-engines we are being forced to use fuels that are less and less gasoline and more vegetables. Who in the would stick carrots in their tank and try to go somewhere and with solar energy failing in every measurable way, why would millions of dollars be pushed into companies owned by foreigners to replace one of our biggest natural resources? Greenhouse gases [CO_2 in particular] became the

14

villains of the 21st century as they pushed us closer to destruction. Instead of Methane which has been increasing in our atmosphere at 50 times the density of Carbon Dioxide, the government focused on eliminating burning and CO_2. I suppose they might have wanted to eliminate farting as well, but CO_2 emissions were easier to tax. Don't even think about describing some type of butt monitor. I for one will not wear one; after all, the Nobel Prize winning climatologist of 2008, Al Gore, didn't say anything about methane so I'm staying quiet. Darn, I can't stay quiet as it has something to do with the data in this book. I have to tell you Mr. Gore in his massive jet fuel burning private air taxi did say something that may have a level of truth when he told us *the Earth is dying because of man.* Certainly his whole burning carbon fuels being the culprit was all a lie, but man is not innocent of some of the weird temperature rises of late.

Let me say one more thing before we get starting in our investigation and that is in the building of plants for fuel. To make them burn better, scientists found a way to genetically modify the DNA of various crops to speed up production for this horrible fuel substitute. The scientists got so excited about changing plant DNA, almost everything was changed. They changed corn so that no horn worms wanted to eat it, and they changed wheat to mature faster. They modified other plants to thrive on the salty environment of desert lands. On and on, one after another the Earth's plant life was converted into something different. We eat it and more

15

and more of us are getting cancer as mutated food seems to be associated with mutated people cells. Oops that sounds like another conspiracy, but I'm just softening you up for the details coming up.

We know something is happening, but what it is and how it will affect us needs to be understood without shouting unfounded claims to support some pet project or money scheme. While our temperature has not been affected, greenhouse CO_2 gas SEEMS to be steadily going up for the last 200 years as shown next. While some tell you reasons, no one apparently knows, for sure why it is happening and what effect, if any, this increase will have. As I mentioned we are looking at the wrong gas, the following chart shows Methane is growing much, much faster. By the way, the CO_2 concentration levels slope increase since about 1965 is not altogether accurate, but we will look at that later.

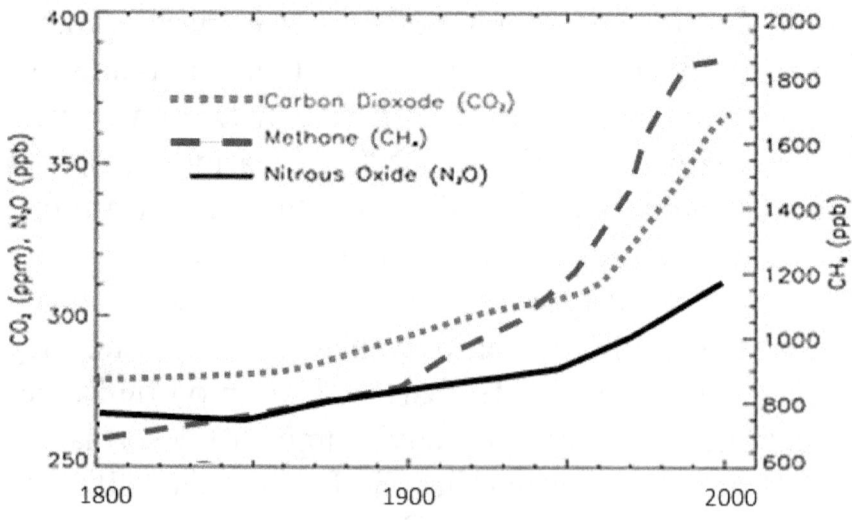

You might be wondering how all of this ties into smoke coming out the back of an airliner and some radio waves, but it will make sense as we go along. First I have to get rid of the notion that most of these chem-trails are benign condensation of moisture. ------ They are not simply moisture. I don't know if you need more proof than my word, but I will tell you that there are large amounts of Aluminum dioxide and Barium Oxide in the effluent from these things and areas where higher concentrations of these "trails" have been seen are experiencing lower vegetation growth and other strangeness. I'll be honest with you, this book does have some conspiracy things in it; I simply could not help myself; especially when the evidence is becoming stronger and stronger. Don't get me wrong. The original 1940s conspiracies probably were unfounded, but sometimes early exaggerations become facts BECAUSE things change. For this book the changes occurred around 1993 as High Frequency Active Auroral Research Program [HAARP] Stations began their seemingly mundane experiments.

Hidden HAARP

If you look on line, this is what you find. Work on the <u>HAARP Station began in 1993 and began trying to heat the ionosphere by 1995</u>. The current working Ionic Research Instrument [IRI] was completed in 2007. As of 2008, HAARP had cost taxpayers $250 million for construction alone and many millions to perform the experiments that are continuously being established. It was reported to be temporarily shut down in 2013 and then in 2014, it was announced that the HAARP program would be permanently shut down later in the year. Now they have extended the shutdown until 2016. Does this seem odd to you????

The since the completion of the IRI the HAARP project can direct a 1.5 Gigawatt signal in the 2.8–10 MHz, HF (high-frequency) band, into the ionosphere that is modulated at extremely low frequencies. [It is this modulation we will look at further.] The signal bounces and part of it comes back to a huge array of HF antennas. [33 acres of the things] According to the HAARP team, this whole thing was to look are the Aurora Borealis and how the natural ionosphere [between 50 and 500 miles up] affects radio signals so they could make GPS more accurate and be able to talk to submarines better. There was also some mention of mapping the mineral content of the terrestrial subsurface, and finding underground complexes or regions in various countries. It should be noted that the ionosphere is high enough to affect how the sun's X-rays and UV rays reach us. Various aspects of HAARP can study all of the main layers of the ionosphere. Some of the main scientific work reported from HAARP includes:

- **Generating very low frequency [VLF] radio waves by modulated heating of the auroral electrojet**. This is, essentially, changing the magnetic field of the entire planet [2004-2013]. Possibly they thought they were simply using the electrojet as a massive radio antenna, but it would have other climate issues.

- **Generating extremely low frequency [ELF] waves** in the 0.1 Hz range. These are next to impossible to produce any other way, so the HAARP antenna Array covers 30 acres. While messing with the magnetic fields of the Earth is dangerous enough, we have very little data on what these ELF frequencies do [2008 to 2014]. Today we are finding that these subsonic wavelengths change our brain functions, can cause terror or pain, can make bones grow; can possibly even make you lose control of your bowels. I know how interesting these experiments must be, but in all likelihood they are destroying our planetary ecosystem.

- **Generating whistler-mode VLF signals** that enter the magnetosphere and propagate to the other hemisphere. This is essentially trying <u>to make lightning</u> [2008-2013]. I'm not going into how scientists thought lightning created life in complex sugars millions of years ago, but what we should worry about here is the massive magnetic field generated by lightning discharges. To increase them will have consequence.

- **Testing methods for increased radio reception** using the Schumann Resonance. [8 Hz-associated with the time it takes for a radio wave to go around the world.] This is scary enough, but now they have been hearing a 0.9 Hz signal called the Alfven Resonance and no one knows what it is as the Earth seems to be groaning on its own when they do this experiment. That can't be good.

- **Producing high density plasma clouds** in Earth's upper atmosphere [2013-2015]. These were made to simulate massive mirrors presumably to reflect radio transmissions, but they also reflected solar energy and captured same.

- **Bouncing HF signals off metalized clouds** placed in an array that resonates at the transmission could allow low range transmission to anywhere. The arrays of metalized clouds could easily and quickly be put into place by having these things called chem-trails be placed in patterns. Unfortunately, the reflective nature of these metalized clouds causes other problems. Ground energy is reflected in the same way capturing heat in between the reflector and the ground causing an artificial high pressure area. ---Wait a minute if we can put the high pressure area on top of a hurricane, we could control weather!----It has not turned out to be that simple.

Confession- the last experiment was not identified anywhere, but I will show some of the effects that show this type of experiment is being done.

Our Ionosphere heating station is located in Alaska and outputs about 1.5 Giga-watts [GW] of energy. In Norway, the European Incoherent Scatter Scientific Association [EISCAT] is still another ionospheric heating facility, capable of transmitting over 1 GW. Three others also continue experiments but are not nearly as powerful. In Sweden we find the HISCAT one, but it has limited use. In Russia their similar center, the "Sur Ionospheric Heating Facility" only outputs 200 Megawatts. In Puerto Rico we find another Ionosphere heating facility called the Arecibo Observatory and still another small one is found in Alaska near the HAARP monster. Let me show you why the Ionosphere heaters are where they are.

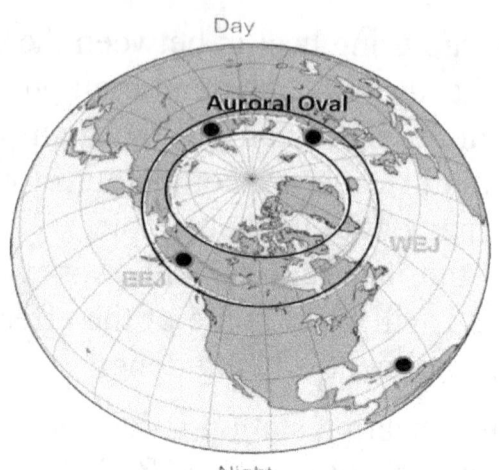

As I mentioned before the region known as the Auroral Oval not only make the Northern Lights, but it also is the easiest place to put energy into the magnetosphere. We'll talk about our Puerto Rico facility a little bit later. Hopefully you can appreciate that these research centers have a very useful side. Unfortunately they are messing with very dangerous things that could destroy our ecosystem. We will look at some of this later, but I just wanted you to know a number of governments are heating our ionosphere with some dangerous electromagnetic waves without enough regard for the potential dangers.

Global Warming

I don't claim to know all details of Global temperature stability or lack thereof, but I surely can show others where the current claims are unfounded, or in some cases extremely odd as if someone was manipulating with the weather. We need to keep an open mind about all this as our Earth is a delicate and important part of our existence and if there is something we can do to protect it we should. Here are a few of the things we hear. Sometimes it is because masterfully established conspiracies and other times simply as directed cautions to get us to buy solar panels or whatever. Later, I will get into these and others in detail but right now let's just get a reasonable list. I'll try not to comment too much.

Runaway CO_2 and Methane levels are going to destroy us Lie- Before a couple hundred years ago CO_2, NO_2, and Methane levels remained more constant than the last 200 year. Something may have happened in the 18th century to trigger this extended release, or the solar system may have shifted, or something bad as shown in the graph to the left. To the right we see the temperature and CO_2 levels over the past 20 thousand years. Just as scary as scary can be. The CO_2 levels are destroying the earth.----so it seems.

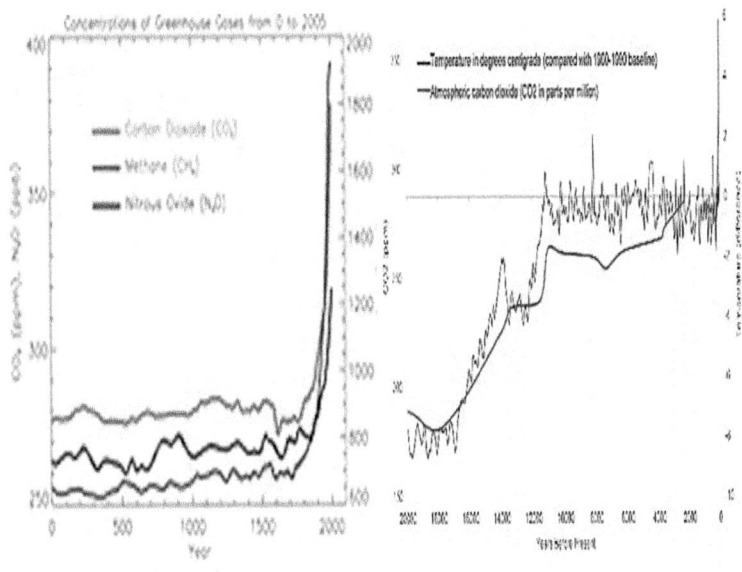

I think I'll wait a little before I get into the subtrafuge. Let me just give you a sneak peak. Those prevous graphs were made from a group of unrelated charts. The main ones of concern are the CO_2 levels packed into ice core samples from Antartica [up through 1996] and the aerosol CO_2 from Hawaii. The sharpe rise at the end of both of these charts is from aerosol data and the first parts are both from Ice Core samples. Just about everyone in the environmental business KNOWs that only a tiny portion of aerosol gases including CO_2 will be trapped in the Ice, one can quickly see the dread has been amplified by some pretty evil people. Below is is a much more accurate set of graphs.

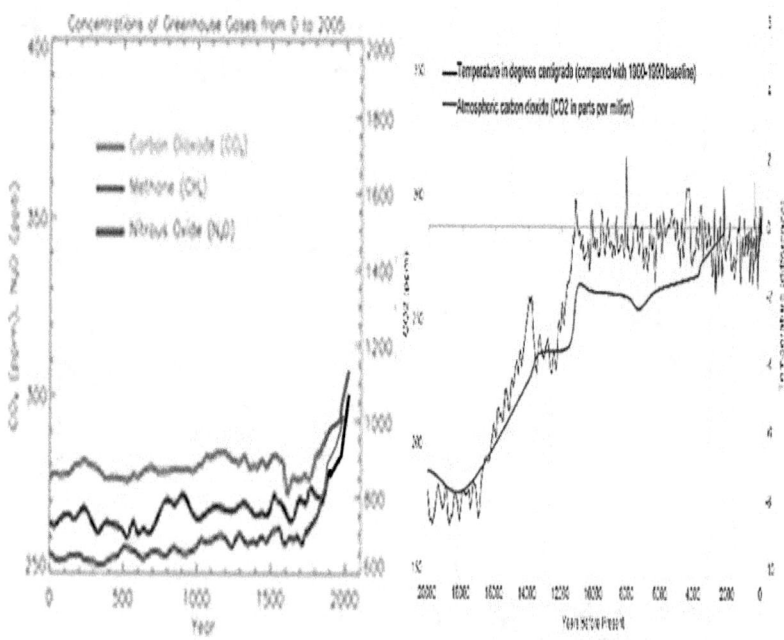

I'll expain all this later and show you just what they did. It is pretty horrible and it was NO ACCIDENT. This does not mean various particulates and trace gases are not increasing. We will look at them and try to determine what is going on without all the sky is falling claims.

Global Temperatures are [supposedly] Skyrocketing Lie- You here, it is "well known" that every year it gets hotter. Never mind the snow blizzards and such, just look at the graphs. This was "compiled by University of East Anglia. It shows about ½ degree over a 160 years, but it looks really bad. Later we will see that the data was doctored to make this graph, but people still are being shown this to get them to support solar and wind power industries. The current temperatures are higher than they were in 1980 but not much higher than they were in the 1940 when temperatures were much higher than on these "official" charts.

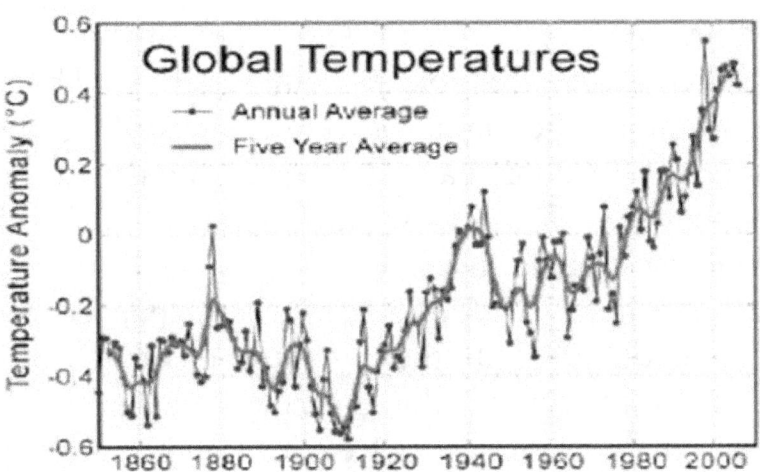

The entire chart has been masterfully altered substantially and no one went to prison even after the act was found out and they had to admit the attempt at driving fear into the entire world. If we look at the raw

UAH satelite data for the tropics and for the globe and compare them 2 Nino 3.4 feed, we find something substantially different. What we see is that in March 1998 we were either 0.9 or 1.3 degrees higher that nominal and in September of 2015 were had dropped to between 0.5 and 0.7 degrees above nominal making a relatively flat temperature model.

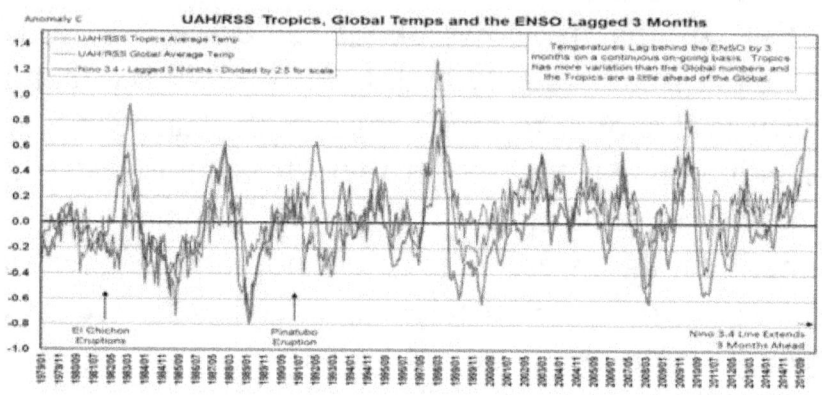

Let me just say I don't know how they determine weather in the first place from satellites. The Image below shows the major temperature recording satellaite data. One would have to be a genius to provide temperature changes on the order of 0.05 degree differences being used today to support the gloabal warming frenzy. Also look at what happened between 1970 until 1985 all the satellites recorded the same thing. Don't worry if this doesn't make sense, We find that the National Oceanic and Atmospheric Administration [NOAA] produces very accurate graphs.

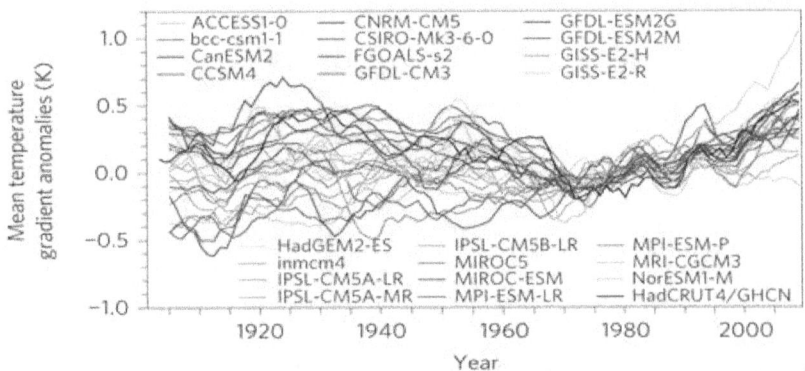

This isn't the whole picture, but right now, let me go on or I'll never get through the introduction.

Coal Is Our Enemy Lie- As I mentioned this is one of the major conspiracies, Noted climatologist James Hansen reported the following: *"Coal is the single greatest threat to civilization and all life* on our *planet."* From his HORRIBLE statement, <u>our government has stopped 150 proposed coal plants and "retired" 110 existing coal plants</u>. Now the US Environmental Protection Agency [EPA] passed new clean air regulations that will cause the rest of the coal plants to shut down and it is virtually impossible for new ones to be built forcing our valuable resource to be thrown away like garbage. Of course CO_2 has little to do with heat storage of the earth and everyone studying atmospherics knows this as well as the largest energy absorber in our atmosphere is water vapor which captures well over a thousand times as much heat energy. [Coal use in the U.S. has plunged 13 percent in the last six years as those owning stock in natural gas cheer and renewable energy sources are bankrolled as shown below. [Numbers are in millions of dollars.]

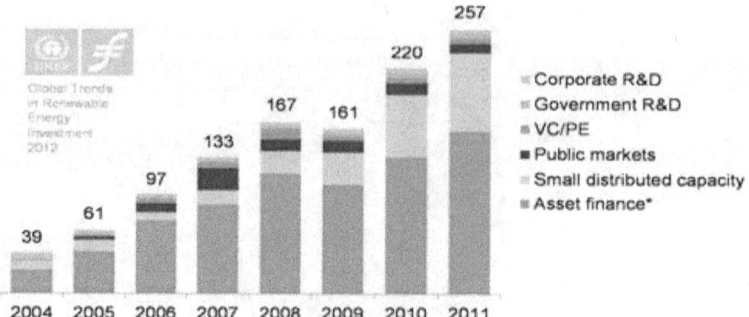

More About Coal- Coal combustion creates 40 percent of electricity worldwide but also is [*supposedly*] responsible for *"30 percent of total anthropogenic [Man-caused] carbon dioxide emissions worldwide, and 72 percent of CO_2 emissions from global power generation"*. As we MUST BE causing global warming, we can CERTAINLY see renewable energy sources like wind or solar should be more cost-effective in the long run. [Just kidding] Quickly, the United States made plans to eliminate nasty Coal and its working as shown below. Don't worry about those in the coal industries or the entire economy of the Coal States, they all can find other work and don't worry that the cost of electricity will skyrocket as soon as coal is not an alternative. A more correct truth will amaze you.

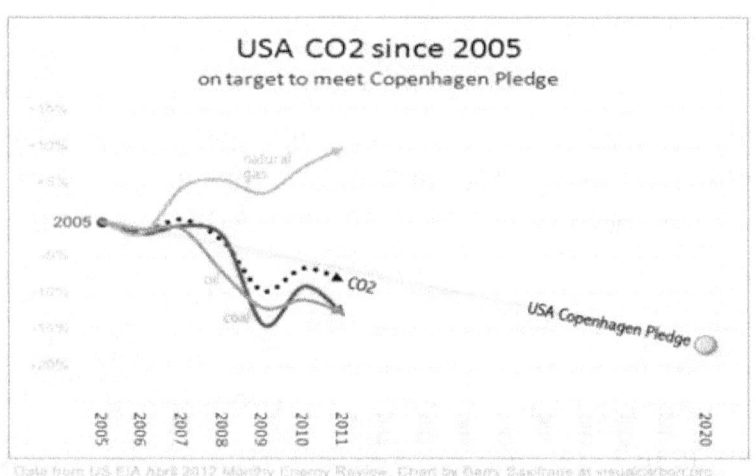

USA CO2 since 2005
on target to meet Copenhagen Pledge

Data from US EIA April 2012 Monthly Energy Review. Chart by Barry Saxifrage at visualcarbon.org

Melting Ice Caps are going to destroy us Lie- The melting ice has caused sea levels in the north to rise. For the first time in hundreds of years, ships can pass through the fabled Northwest Passage above North America. Over 100 million people living in coastal regions will be displaced by just a one-yard rise in sea levels. This is one we are going to look at more carefully as there is an oddness about it.

Sea Level 1900-1980

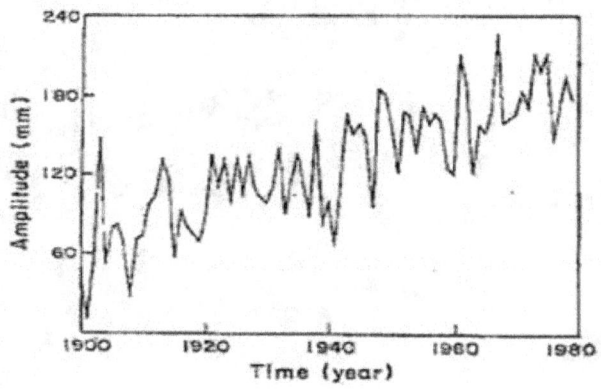

29

Loss of Glacier area in Alaska Truth- One thing that is for sure, the size of many glaciers in Alaska are getting smaller as the mean temperature in Alaska is higher. Let's concentrate on the dates again. This reality is something that our HAARP experiments can help us with. The famous McCarty Glacier picture is pasted everywhere [First row 1909 and 2004]. Muir Glacier is shown in 1895 and 2005 [2nd row], the Mendenhall Glacier shown in 1894 and then in 2008, and finally the Portage Glacier from 1914 and then in 2004.

While it is known that Northern Alaska is getting warmer, just as many areas are getting cooler and lets look at the dates as massive loss of glaciers happened between 2003 and 2009 when just about everything else seems to have happened. Somehow NOAA couldn't even think that the HAARP blasting was causing something bad to happen.

- **Yes;** they are using ELF waves that have been shown to be extrememly dangerous.
- **Yes;** the drive massive electrical pulses into our magnetophere who closely locked magnetic field is the only thing that separates cosmic and Xrays from our earth.
- **Yes;** they are are running weather control experiments, but nothing can go wrong with weather when people are running experiments on it.
- **Yes;** they are making artificial Aurora borealis things and lightning, but that is just child's play.
- **Yes;** there are anomalous high pressure areas that build around the site, but how can a little high pressure area hurt anything.
- **Yes;** they have admitted to tampering with the weather, but we should not listen to them as they may have been coerced.

Methane and Flatulence- We are told cows produce more methane than the oil industry does: 20 percent of U.S. methane gas emissions is produced by farmed cattle burps and farts. [No telling what dinosaur flatulence did.] Sorry about that, but this is another area we need to look at very closely as the good old HAARP and Chem-trails seem to be helping this problem get worse.

Drought- They tell us, because of global warming, of the land in California, 99.84 percent is experiencing drought. Unfortunately the drought part is correct. Simply saying global warming is not the whole story. Please remember this problem because we will be getting back to it.

High Cost of Global Warming- Climate change costs the U.S. over $100 billion each year. The problem is, we are not having a positive effect on reducing those nasty greenhouse gasses.

Global investment in renewable power and fuels increased 17% to a new record of $257 billion in 2011. Developing economies made up 35% of this total investment, compared to 65% for developed economies. The US had a 57% leap in its outlays to $51 billion. India increased its efforts by 62% to $12 billion in 2011 alone.

Loss in our Seas Lie-The world lost about 16 percent of all coral reefs in 1998, the second hottest year on record due to greenhouse gasses [according to doctored data]. We have caused this!! Since the beginning of the Industrial Revolution, the acidity of surface ocean waters has increased by about 30 percent. Massive Thermal Runaway will hit us very soon as ozone is disappearing, our surface temperature is unlivable and acid in the oceans is killing everything as shown below in the NOAA published graphics. The problem is they are not showing the truth. Don't get me wrong, we do have a problem, but we need to fight the right battles.

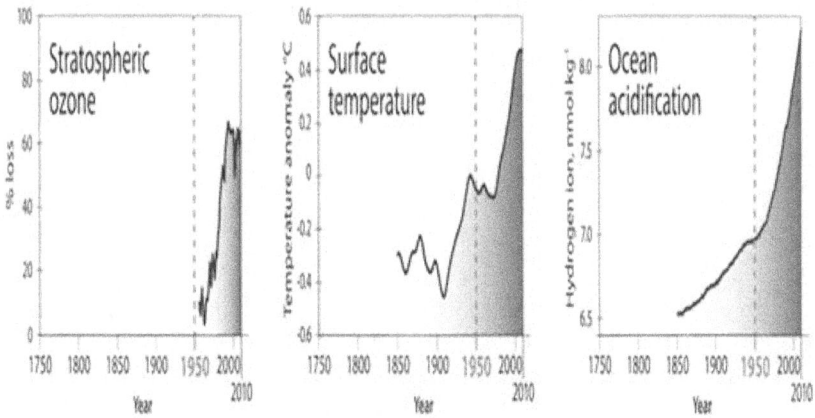

Antarctic Ice is Leaving Lie

Like the Arctic, we are told Antarctica is losing its Ice. They even have a graph showing that at the bottom of South America Ice is getting les, but most of Antarctica is getting more and more ice as this is just another lie. The NOAA graph on the right shows huge reductions in Antarctic mass over a 6 year period.

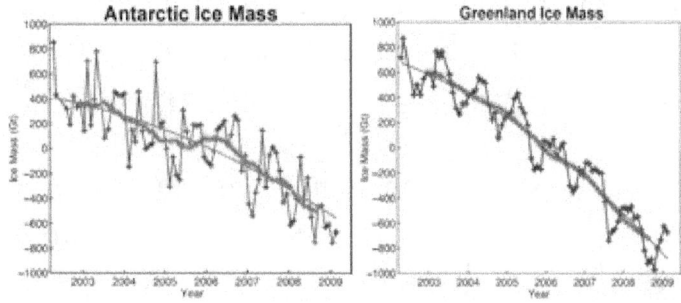

The graph on the preceding right shows almost the same catastrophe in Greenland and the graph below show the ocean heat is rising as fast as a fish can swim.

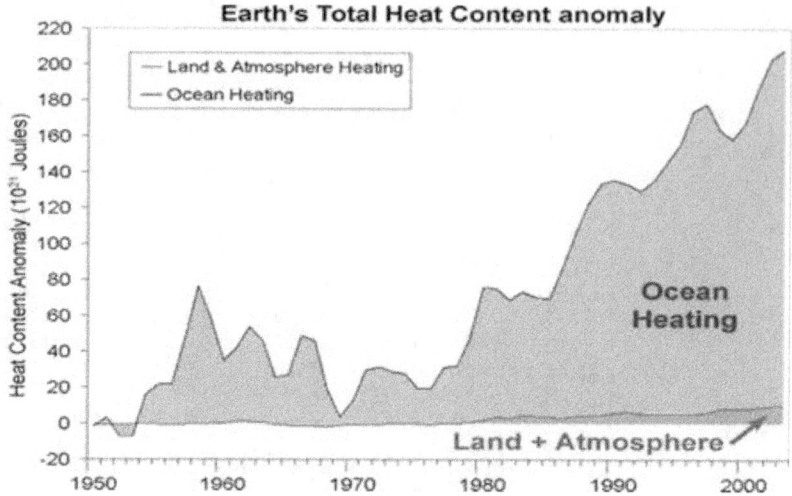

Wait just a minute!!! This doesn't sound right. Satellites show the opposite. "Fortunately", some of this stuff is exaggerated, manipulated, falsified, introduced by unqualified quasi-scientists, and many details are buried to keep the world populations supporting the green-gas elimination industries and even to get a Nobel Peace Prize or two. Unfortunately there are areas that are heating up on our planet, but oddly, it not the entire world as those wanting to divert issues and build greenhouse gas empires. If we look at the following graph we see something very peculiar. While there seems to be a slight rise is the each temperature since 1900 [which we will look at more closely, there is absolutely a more massive temperature in the northern hemisphere that the entire bottom of the earth.

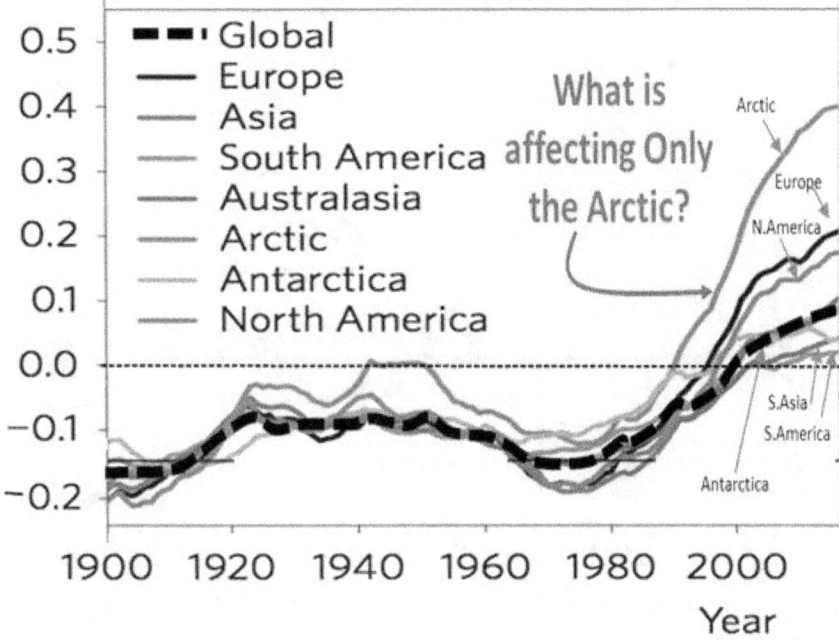

Average 40-year rate of change: CMIP5 RCP4.5 scenario

To make this a little clearer, I have shifted the nominal rise out of the chart and clumped the increase by latitude. See the next revealing graph as everything seems to be good until about 1995 or so.

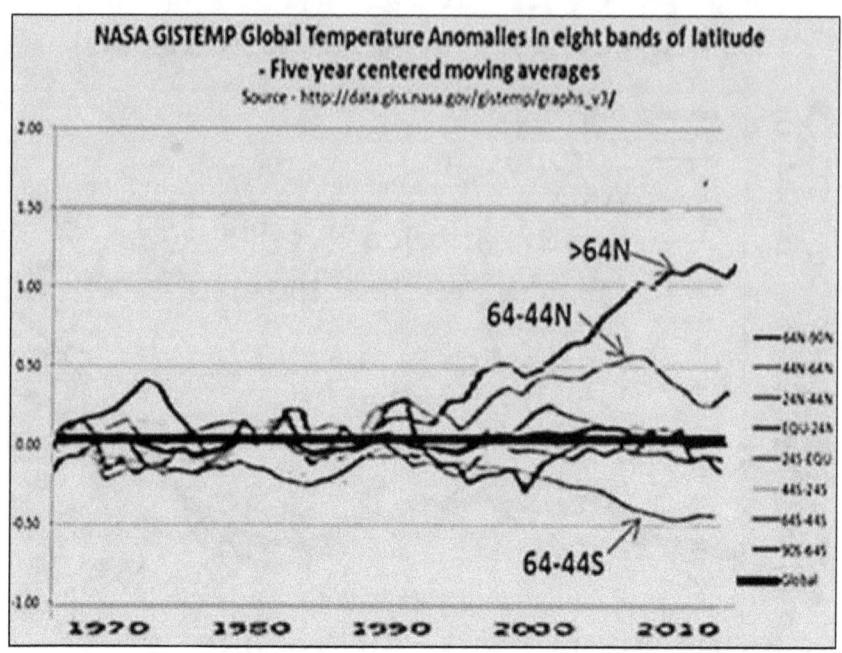

NASA GISTEMP Global Temperature Anomalies in eight bands of latitude - Five year centered moving averages
Source - http://data.giss.nasa.gov/gistemp/graphs_v3/

This book simply will go through the data and some of the bizarre findings of both chem-trail emissions and HAARP- ELF transmission that are much more serious than putting gasoline in your car. What we find is that there is a good reason the Arctic is melting while the Antarctic is as cold as ever or even colder. I don't want to simply tell you horrors; I want to let you make up your mind. We'll look at previous extinction times and determine what happened. We will look at Ice Cores, Sunspots, Cosmic Rays, the other weather sampling data that has been truncated to make it appear to be bringing us to calamity, destruction models, that help us understand how our Earth behaves and how we have been lied to. Then I will bring in details from some of the experimentation with Ionospheric Heating and how planes filled with AlO_2 and BO_2 factor into this whole mess. What we will find is the following:

We have been lied to over and over and over again and the fear of Global Warming has been greatly expanded with this false data. It appears that some of it is to divert attention from intentional or unintentional experimental results while another use for the false data is to line pockets of the Greenhouse gas Czars.

While you think this world wide lying could not possibly happen, I will show how it is done every day—starting with our children.

We'll find that a huge meteor hit the earth and caused millions of tons of magma to be expelled without our earth atmosphere being affected over the long term. Then we will find that ½ million meteors hit after that and we still didn't turn into a ball of fire like Venus.

Many times in our ancient history, massive civilization rose and fell without much of an impact on our atmosphere. Massive increases and decreases in out atmospherics are cyclic and well out of the control or damage of gasoline engines.

Climatologists tie different types of data together without regard for reason to make horrible claims that expand the fear of the destruction of our planet. They gain notoriety and we gain fear. There may be another reason for the deception that will not make us feel any better, but it may explain a few things.

Easing Fear of CO$_2$

While I'm going to describe some of the key elements of global heating and some of the ramifications, what I don't want to do is enhance fear of something that is not happening. That being said something is happening and we desperately need to understand more about the truth than what is being offered by political groups like NOAA and others quasi-scientific organizations bent on making their assumptions or what they want to believe more real without addressing or purposely hiding anomalous data that would make it harder for them to provide answers that would satisfy those funding some of the "research" to hide a fairly significant issue.

Cretaceous Extinction Section-To start off, I'm going back to the worst destruction man may have ever witnessed. Yes; I'm talking about the complete destruction of the world called the Cretaceous Extinction. No one was worried about global warming, but there was a catastrophe coming just the same. As we take this second look at our history, please understand the Earth did not melt as billions of people came and went through war and devastation not associated with global warming and the earth has gone through many thermal cycles. In fact they happen every 100 to 120 thousand years whether we are here or not.

Pleistocene Extinction Section -Some of this will seem strange to you as the global warming misdirection has also been used to rewrite history over and over again. Once we get a better understanding of a more probable history, we'll look again at the notion of global warming with our eyes more open and we can more comfortably address HAARP and Chem-Trails. No matter by who or why, what we find is that <u>we have been lied</u> to and there is substantial proof. We will look at why a little, but mostly I want to just give you the data. This next section deals with our earth up until the end of the Pleistocene and into the Holocene before the United States became a nation.

The third section- This deals with the debate concerning true global warming, what manmade elements are being blamed, How the data is manipulated, and the levels vested partners will go to until their ideas are considered true, no matter what they know to be the consequences. In this section we will begin to investigate the ELF Ionosphere heating and Chem-trails. The last section deals with impending changes, what we can do about it, if anything, and how these changes may affect us in the near future. Let me start way back in time as a substantial greenhouse gas producer was beginning to take shape. I'm talking about huge Dinosaurs. Some suggest that the Dinosaurs actually made oil as their bodies decayed. While there is little evidence of that; what I'm talking about here is that Dinosaurs must have farted a lot and produced methane. This was not millions of years ago as some still try to push, but today there is a new timing model.

Not Very Stable

Just how easy is it to fool and entire world into thinking man causes Earth to be unstable? This initial overview doesn't really affect the realism or fantasy of experiments done by Ionospheric heating centers around the world and Global, the dangers of Chem-Trails, the warming by Greenhouse gas, and fear that we should or should not have, but it is an area where evidence has been kept from many to secure some irresponsible course of keeping the status quo no matter how erroneous the information is known to be. The ruse was so inviting, that in general, the entire world was convinced of something that turned out to be wrong. Let's look a somewhat different subject to see how the world population can be fooled in huge groups. I know you are going not going to believe me initially but nuclear decay [all that Carbon 14, Uranium, Nitrogen, and all type of isotope testing] as a form of "accurate and consistent dating calibration" has been KNOWN to be a lie. For some time now, the entire science community has been struggling with huge issues in their previously established timing baseline. You have been told dinosaurs died 65 million years ago and the beginning of the Mesozoic Era was 300 Million years ago. You were told, over and over and over again. It was proven to you that was truth by telling you lead, potassium, and even carbon isotopes decayed at a set rate; just see how much of an early isotope is left and read the date. Besides; the dinosaurs are buried underground and turned into stone so there had to be a long time for that to happen.

Ancient Earth

While the earth is ancient, it is definitely not as old as has been told to you. Many of geologists today still tell you that radiometric dating has narrowed the age of Earth to about 4.5 billion years, give or take a couple of percent. We now know that the dating method is inaccurate. Scientists not wanting to correct the error, promote what can be called vain truth [truth they want to believe or propose for some ulterior motive] Unfortunately, the Earth and everything in it is much younger and so are the characteristic stabilities of the planets in our Solar System, including Earth. Researchers at Purdue and Stanford have found evidence that radioactive decay rates are not constant at all. On December 13, 2006, a magnificent solar flare flung radiation and solar particles toward Earth. Measuring the decay rate of manganese-54 during the flare proved to be very interesting as the decay rate dropped during the time of the radiation fallout. It was determined that solar neutrinos zipped through space and affected Mn-54's decay rates used in the experiment. Just think about this. They were testing a single solar flare event and the change was significant. The sun has these things all the time. It was also found that the decay rates of silicon-32 and radium-226 showed seasonal variation, according to data collected at Brookhaven National Laboratory on Long Island and the Federal Physical and Technical Institute in Germany. This error was just the material sitting there with almost no outside interference. Wood buried in igneous rock in Queensland Australia has been dated to 40 thousand years, while the basalt around it dated to 45 million years. Both dating subjects should have given the same date, since the igneous rock was formed at the same time the wood was buried. Many of the "data-ologists" don't tell you about major errors like this.

Lava Errors

Excess argon-36 was found in three out of 26 lava flows in recent times. So Argon/argon testing would show a much older date that actually was "KNOWN". This is believed to be because there was too much of the argon-36 in the first place. In the Grand Canyon lava flow testing showed lower levels of lava were younger than the top layers. At different volcano sites, that had eruption in 1949, 1954 and 1975. The same thing was noted. Geochron Laboratories of Cambridge, Massachusetts dated these samples. Even though the oldest of these samples are just over sixty-years old, the lab tests provided ages that ranged from 270,000 years to 3.5 million years old. Additionally, we go to Mt. St. Helens and its eruptions in the 1980's. Samples there gave old ages in the range of 300,000 to 2.7 million years. Hopefully, you are beginning to see that we know less about how old we are than you believed before reading this. If neutrinos from a single solar flare can make things look older, what if the entire Earth was closer to the sun? I know that sounds odd, so just keep it in the back of your mind right now as we try to find some standard for dating. Other methods had to be employed to determine how everything should be timed, but classroom information was not changed. That would confuse the students. I'm going to prove to you how you have been lied to. This will give you a better understanding of the lengths some will go to when they believe something, no matter what the evidence shows.

Standard Geological Timeline

Era/Period/Epoch	Time (M yrs. ago)	Time (T yrs. ago)
Archaeozoic Period	5000-1500	50,000-3000
Proterozoic Period	1500-545	3000-1000
Cambrian period	550-500	1000-900
Ordovician period	500-440	900-800
Silurian period	440-410	800-700
Devonian period	410-365	700-600
Carboniferous	365-300	600-500
Permian period	300-250	500-400
Triassic period	250-212	400-300
Jurassic period	212-145	300-200
Cretaceous period	145-65	200-100
Tertiary period	65-1.8	100-40
Pleistocene period	1.8-0.01	40-10
Holocene period	0.01-0	10-00

The middle listing of dates is the "STANDARD" that had been presented in our classrooms, while the last column shows a somewhat closer, more accurate time line that has been verified by non-nuclear decay methods. Even with the mountain of evidence showing how nuclear decay cannot be used, the middle timing is still heralded as the master in many schools and books being used to teach our children <u>without basis</u>. I know it is difficult to believe historians, scientists and teachers would keep these things from you, and information like how greenhouse gas has little effect on our planet while cloud cover and sunspot cycles are killers, so let me tell you a little more.

Stratigraphic Positioning

Besides nuclear decay, the main way scientists used to determine "age" was by stratigraphic positioning. This is the determination of age by position, depth, and material consistency. MANY TIMES this is <u>the only method</u> for cross comparison that was thought to be reasonable for confirmation of radioactive decay. Scientists simply determine the depth of objects, or features near the object, or number of lava flows, or similar geologic characteristics and use the depth as a time gage. Scientists have been using this for a long time and assuring us this was a great way to time when the dinosaurs were here when, all of a sudden, there were trees found that were going the wrong way. The next set of pictures shows some of the unfortunate trees that must have died repeatedly to be deposited perpendicular through all of the stratigraphic lines.

Some try to state the trees were millions of years old or they simply fossilized while standing for MILLIONS of years as the ground built up around them. [20 points on the BUNK meter!]

Let me ask you something. If neutrinos from a single solar flare can make things look vastly older, what if the entire Earth was closer to the sun a couple hundred thousand years ago? I know that sounds odd, so just keep it in the back of your mind right now. Right now, I'm going to provide you with a more logical way to time the beginning of the Triassic Period as our planet rotation was [and still is] not stable. That is where Ice samples come in.

Ice Core Cycle Dating

Although the task is tedious, ice can be examined just like tree rings. Each summer ice changes its consistency. H_2O (16) is more concentrated in the summer while H_2O (18) is more concentrated in the winter. This gives us indication to the level of CO_2 which in turn allows us to understand something about the temperature levels. As the yearly cycle has freezing and thawing, ice consistency varies each day, seasonally, and yearly, independent of the Earth axis and other critical elements. Anyway, scientists around the world started boring holes in ice.

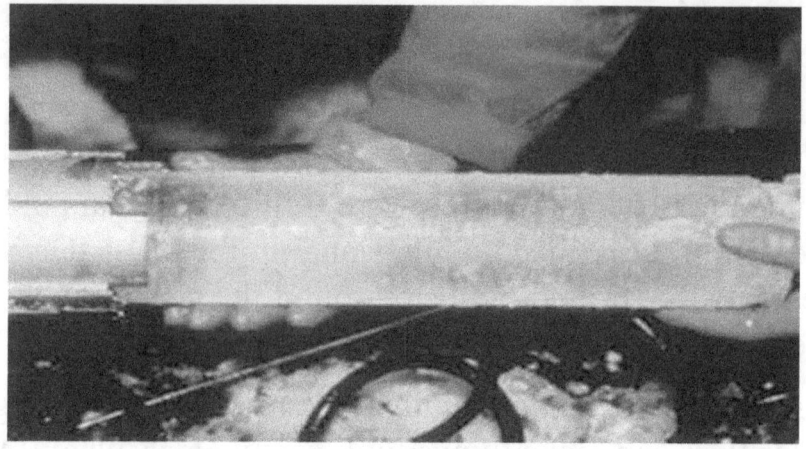

The most coring is done in Greenland and Antarctica. A sample is shown above and part of the ice coring record is shown below.

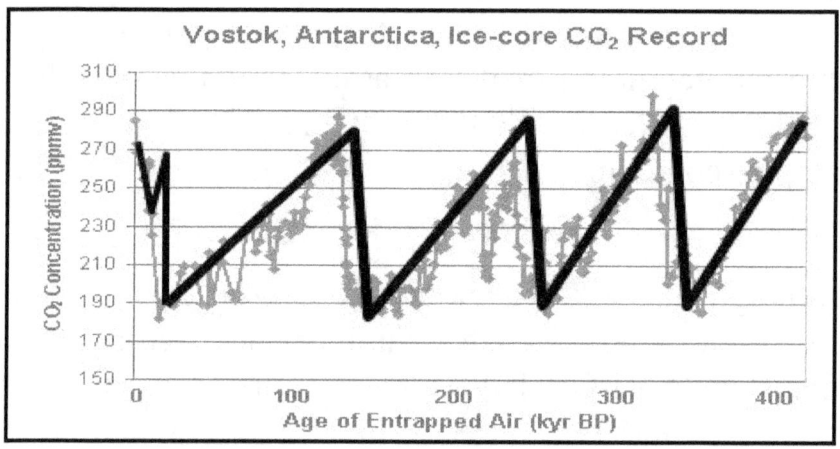

If you look closely you can see that about every 120 years there is a massive change in the environment. One would think these "corrections would cause mass extinctions. This was found at both Antarctican and Greenland Ice cores and the dating is by daily, seasonal, and yearly changes rather than nuclear decay. Bah humbug! You say! Well, what if we see confirmation?

Hawaii Hotspot Track Dating

Hawaii is not a tiny group of islands, but instead is an indicator of where the Earth magma has a hotspot. As the crust moves differently than the stuff below, the hotspot relative to the crust moves and each time the hotspot burns through another piece of crust, a volcano erupts which seals off the area after a time and an island is made for a few thousand years. This travelling hotspot known as Hawaii is show next. The descriptions provided shows what was happening along the way. Because the hotspot moves <u>perpendicular to the axis of the Earth</u> we also know how the earth was spinning as shown by the lines in the first graphic below, but the actual timing is not described here. I placed some general times in the second graphic, but let's see if they make sense.

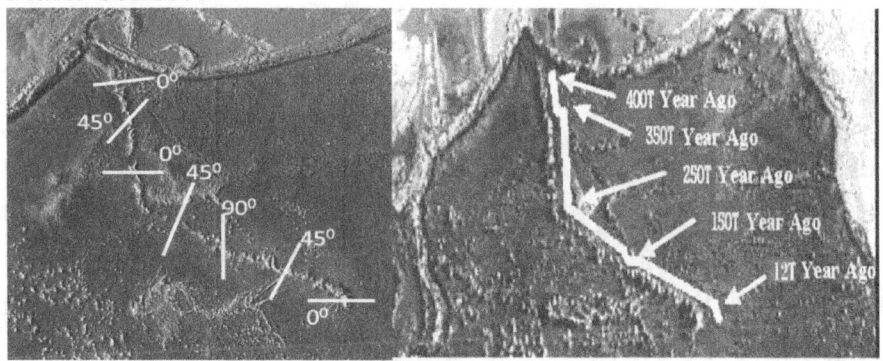

Let's compare the Earth shifts with the Ice core data. Man-oh-man; it seems they match. I think you still believe in nuclear decay so we will look farther.

47

Hopefully you will begin to believe that the Earth environment changes substantially on its own whether people are driving in cars or not.

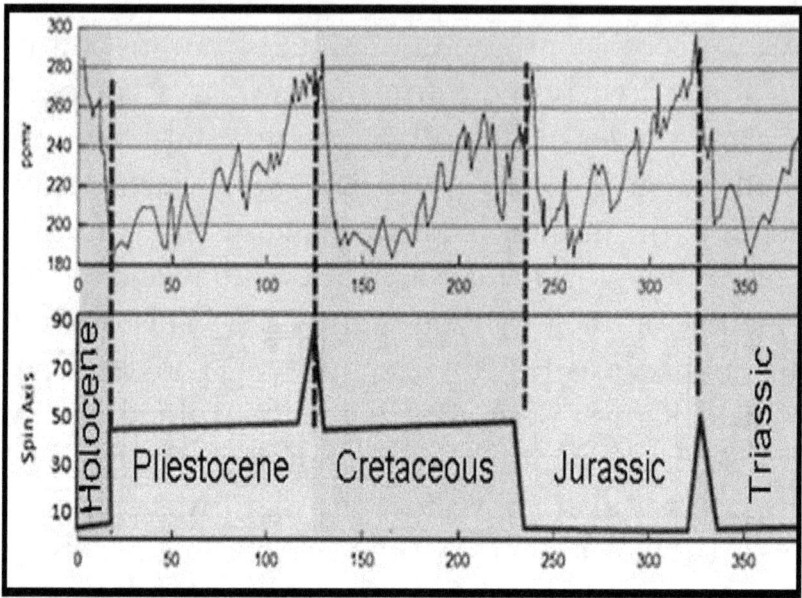

The Atlantic Ocean is getting wider about an inch a year, averaged worldwide. While the building of the great mountains has little to do with the normal tectonic plate "drift" We can pretty accurately measure the widening ocean in various ways including measuring distances between matched magnetic landmarks on either side of a widening gap on the ocean floor. The old theory indicated that 180 million years ago the continent Pangea began splitting apart and has been drifting ever since. In so doing, the landmasses of the Western and Eastern hemispheres separated and opened the Atlantic Ocean basin today, but that was not the whole story.

Plate Tectonics

Plate tectonics tells us the outer hard crust of Earth consists actually of a dozen or so distinct, hard plates that drift individually on hot, deformable rock. An unequal distribution of heat within Earth moves the plates. The boundary between the plates forming the Atlantic Ocean is smack down the middle along the Mid-Atlantic Ridge, shown as the hashed line in the figure below. The ridge is where we must look to find a widening gap, which accounts for the widening ocean.

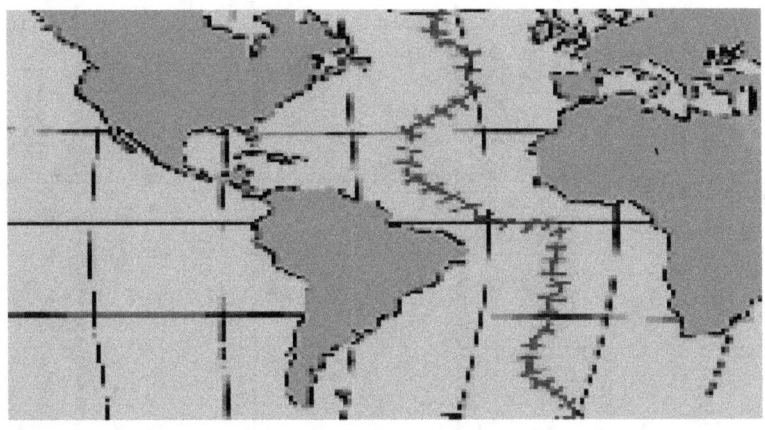

That is where we measure the rate of separation. Where the plates separated, white-hot soft mantle oozed up from great depths within the Earth to fill the gap. The molten rock cooled slowly into new slivers of sea floor. This happened over and over again through the eons. That's how the Atlantic Ocean widened-by a spreading sea floor. Iron-rich rock has a peculiar property; heat it above its curie point of 580 degrees Centigrade and it loses its magnetism. When it cools the rock gets re-magnetized in the direction of the existing Earth's magnetic field. So it's a magnet with the poles aligning with the poles of the Earth at the time of the cooling. The neat thing about this is: the magnetic field of the rock, once cooled, stays frozen in this orientation. It becomes a record of the Earth's field at the time of its cooling. The first graph below shows how the magnetic field has changed over time. Certainly we cannot get an actual time, but a relative timing is very good. What if I told you this matched up exactly with the Ice Core and hotspot data.

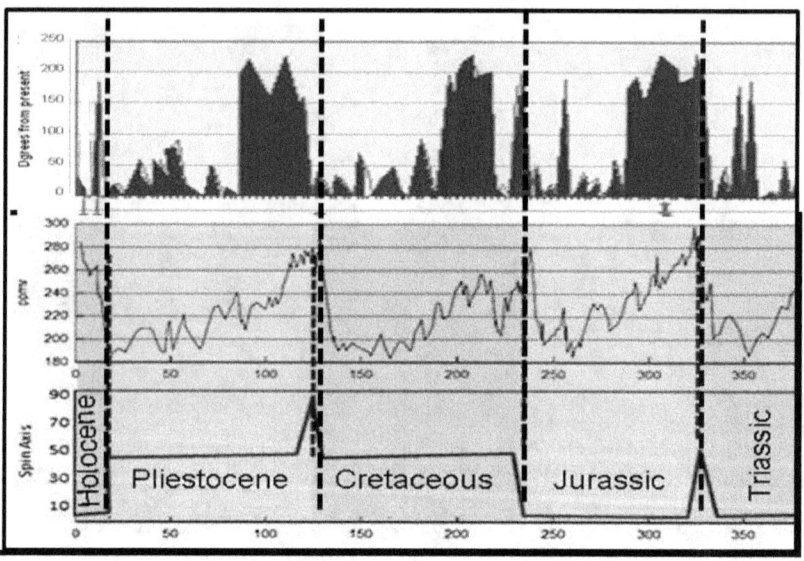

The graphic below tries to show some possible major earth "settling" points and general information about those spin axes. For instance, notice that the earth spin goes along the east coast of the United States 12 thousand years ago and shifts to where it is today very shortly after than time. You would think when that sort of thing happened the atmosphere would have been in trouble and we would experience greenhouse gas runaway, but we did not so some scientists are trying to keep it a secret.

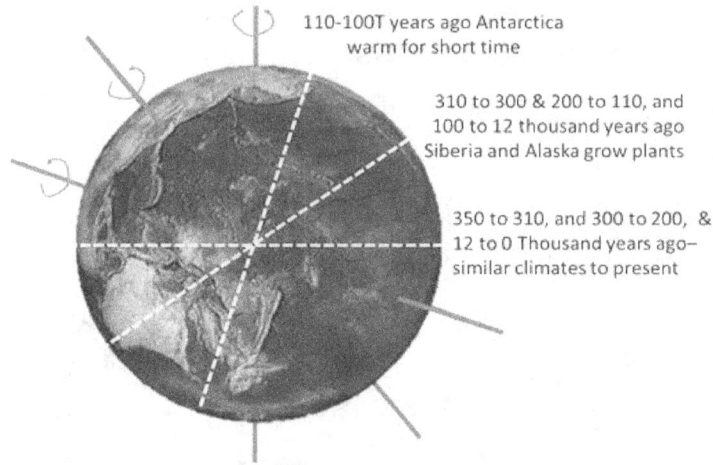

110-100T years ago Antarctica warm for short time

310 to 300 & 200 to 110, and 100 to 12 thousand years ago Siberia and Alaska grow plants

350 to 310, and 300 to 200, & 12 to 0 Thousand years ago— similar climates to present

Marine Isotope Stage [MIS] Dating

Some people may still be reluctant to give up what the schools have been preaching so very long, so I thought I would bring out one last attempt at presenting sanity. Large numbers of scientists around the globe are doing Marine Isotope Stage timing by digging in dirt. It seems looking at the levels of Oxygen 18 shows how hot or cold a point is in time while checking relative Oxygen 18 isotopes in Calcite [which just happens to be the main ingredient in seashells], one can tell just how many of the things were here during each period. Checking around the globe has given us a good map about climate and number of seashell, which correlates to number of animals in general so it is easy to see where extinction periods are. Guess where they line up? Time's up! They are an almost exact match as shown below. MIS levels are shown next above the ice core samples, the hotspot data and the magnetic field shift data Massive drops in O_{18} mean massive drops in sea shells and all other life. Notice there is no extinction period between the Tertiary and Pleistocene Ages marked by Cro-Magnon appearing. Please say you see a comparison.

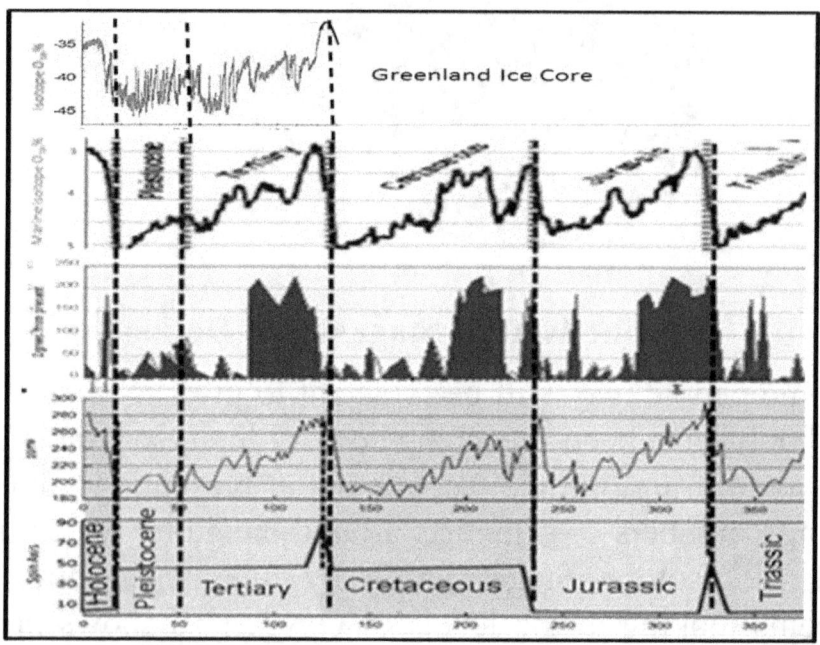

While nothing except for nuclear decay timing "sometimes" showing extremely long times and everything else shows cross comparative affirmation of a realistic dating which shows each extinction , each Ice Age, each shift on our axis, the expansion of the Atlantic Ocean, and major wars--- has been hidden from most around the world. The lie was easy just lie the lie that is going on today about these chem-trails and the damage done by the HAARP device.

Why Am I Putting You Through All of This???

People, scientists and governments all lie. I don't mean in a conspiracy way. What I'm talking about now is lying done for fame, money, payoff for favor, some mistaken philosophy, fear of reticule after pushing a mistaken philosophy, or just because they are afraid and feel obligated to present a special issue they feel will save the world no matter what the cost. The manufactured global warming misdirection is a little of all of them.

Global Misdirection

When someone tries to tell about problems with these chem-trail things and the exuberant scientist heating of our Ionosphere with exotic subsonic vibrations at the massive HAARP laboratory it sounds like a conspiracy theory and I suppose one could identify this detail in that light, but the data is still going to show problems in our atmosphere have nothing to do with CO_2. The timing, area of change, level of disaster, and characteristics of the experimentation show closely associated destruction and change when compared to

HAARP testing and Chem-trail blanketing. I cannot guarantee that our environment is not being adversely affected by the worst green-gas on the planets [H_2O] but for some reason no one wants to talk about this danger, especially as it relates to attempts at weather control. I know the dating I presented is not what you were told and I know the idea of using nuclear decay to date things has made a comfortable geologic record, but the pieces don't fit. The mutation of mankind doesn't fit and new discoveries of non-fossilized dinosaurs don't fit. Still the entire world is still being told the Earth timing is 1000 times as long as all other timing resources tell us. By making our Earth seem much older and extinctions much more distant in time, scientists can tell us that evolution can turn various sugars left on a stump and turn them into DNA and lighting took the very complex sugars and drove life into them. It kind of makes sense if there really was 4.5 billion years. If someone can do that, telling the world that driving a car is destroying our planet is a piece of cake and it diverts questions from other things being done.

Recent Lies Cover Up Real Warming

Sorry, but I simply have got to vent here before a have to throw up. Just like the quasi-scientists are still pushing nuclear decay timing, there is something far worse being perpetrated that is still not about the Chem-trails. I'll bring it up again later, but I want to show you something that is horrible. The following is supposed to be Ice core samples from Antarctica. If you look on the NOAA site, you will see something similar. I circled the thing that is different on the very right side of the charts.

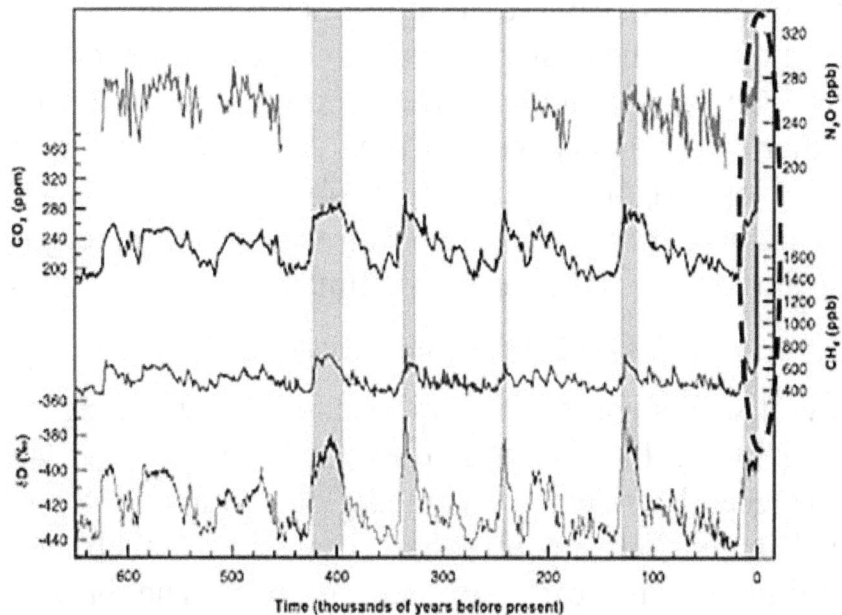

Time (thousands of years before present)

Ice coring has one issue; the top layers of ice are pretty well contaminated so the Antarctican Coring only could date up until about 1970 or so. The Greenland Ice cores had similar issues. I don't know who initiated this ruse but <u>NOAA has pushed it hard</u>. From 1970 until today, the data is ALL FALSE-MADE UP, ERRONEOUS, A LIE--------. I almost said something I would regret. This was done to make it look like Antarctica had recorded an unbelievably massive increase in CO_2, NO_2, and Methane. Even though the sharp increase depicted is impossible, NOAA and the other fake groups continuously use the faked charts----AND THEY KNOW IT IS A LIE!! Just like convincing the entire world that nuclear decay timing proves survival of the fittest style evolution, the making of oil by billions of dinosaurs, this comes along to prove jet fuel was destroying our planet and burning coal would bring us destruction. The hysteria seems to be getting worse and worse and people ignore what is going on right in front of them. Originally the warming earth would raise the sea level and there were many, many charts showing the massive increase in the ocean waters, but here goes the stupid earth going to flood climatologist.

CO_2 Making the Ocean Sink

Now it seems that there is so much global warming that it is slowing the rise of sea levels. From 30 years ago until recently, scientists blamed global warming for sea level increasing. Imagine that! There is so much snow melting that our oceans are getting lower. The STUPID government agency [talking about NOAA] said that from 2002 to 2014, it looked at changes in gravity as well as estimates of glacier melt. The Daily Mail reported the analysis suggests climate variability resulted in an increase of about **3.2 trillion tons of water being stored in land** so the oceans are getting lower. How dumb do they think we are? What really is happening is that sometimes it is good to make the Arctic Ocean warm and other times sanity weighs out and experiments are slowed to allow for a more normal cooling effect.

Almost at the same time, the National Science Foundation estimated that a large part of the Antarctic ice mass appeared to be collapsing, which could cause

56

oceans to rise by "almost 20 feet." It was pointed out, *"There is no evidence of an acceleration of sea level rise, and therefore no evidence of any man-made effect on sea levels. Sea level rise is primarily a local phenomenon related to land subsidence, not CO_2 levels. Therefore, areas with groundwater depletion and land subsidence have much higher rates of relative sea level rise, but this has absolutely nothing to do with man-made CO2."*

Reinforce Mania to Divert the Truth

Even with the Arctic Ocean as the only area getting blistery hot, even with almost not automobiles in the area, the quasi-scientists unite. To show how well people have been bamboozled; even with all of this data showing it is a lie, some two-dozen scientists, [who had written massive reports on how the Earth was dying because of man] with major U.S. universities **[these are the people teaching our children]** urged President Obama, a few months ago, to use RICO laws to **prosecute** opponents who deny mankind is causing catastrophic changes in the climate. In a letter addressed to Obama, the scientists said, *"They appreciate that you are making aggressive and imaginative use of the limited tools available to you in the face of a recalcitrant Congress. One additional tool – recently proposed by Sen. Sheldon Whitehouse – is a RICO investigation of corporations and other organizations that have knowingly deceived the American people about the risks of climate change, as a means to forestall America's response to climate change. If* *corporations in the fossil fuel industry and their supporters are guilty of the misdeeds that have been documented in books and journal articles, it is imperative that these misdeeds be stopped as soon as possible so that America and the world can get on with the critically important business of finding effective ways to restabilize the Earth's climate, before even more lasting damage is done.*" While some of these "scientists" are just stupid, others probably knew about the effects from the experiments done by HAARP. Both are dangerous.

London's Independent newspaper declared at the turn of the millennium, "*Snowfalls are now just a thing of the past.*" The report quoted David Viner, senior research scientist at the Climatic Research Unit of the University of East Anglia, long considered an authoritative resource for global warming research, saying "*snow would soon be a very rare and exciting event in Britain. Children just aren't going to know what snow is,*" he claimed at the time. <u>Oops! I guess the leading researcher didn't know that the United States could turn on and off the apparent Arctic Ocean climate fiasco at will.</u>

Al Gore told an audience in a 2009 speech that "*the entire <u>north polar ice cap during some of the summer months could be completely ice-free</u> within the next five to seven years.*" And his 2006 documentary, "*An Inconvenient Truth,*" *famously predicted <u>increasing temperatures would cause earth's oceans to rise by 20 feet.</u>* Of course it was totally absurd as the United

States could turn off the cooling while still burning more and more coal and gasoline. I wonder if anyone will be convicted of an Obama "RICO ACT" prosecution. These "special" scientists should be so proud. They certainly are selling a lot of books.

Others Get the World Focused on CO_2

A 2013 column by Mark Hertsgaard was headlined: *"The End of the Arctic? Ocean Could be Ice Free by 2015."* He wrote: *"Say goodbye to polar bears and a whole lot of ice. New research suggests the Arctic Ocean could be ice-free by 2015, with devastating consequences for the world."* Al Gore went one better predicting that the *"Polar ice cap may disappear by 2014."* The BBC would not be topped so they said the *"Arctic summers would be ice-free by 2013."* The Sierra Club Canada also said in 2013 that *"the Arctic sea ice would vanish that year"*. Tim Ball, a former University of Winnipeg climatology professor lied and said, *"Global temperatures have been dropping since the turn of the century, prompting the change in terminology from global warming to climate change."*

Here is a tiny amount of data they had to make this determination.

NASA looked at satellite measurements collected over recent years to estimate the rate of sea level rise has slowed by 44 percent form reported 0.016 inches to less than 0.007 inches increase per year between 2004 and 2014. For those worried-----THAT AIN'T MUCH!

59

The sea level rise over the past 200 years "shows no evidence of acceleration, which is necessary to assume a man-made influence.

The National Snow and Ice Data Center in Boulder, Colorado, showing the Arctic ice mass, as of Sept. 7, 2015, was substantially bigger than it was in September 2012.

From 2012-2015, the Arctic ice mass "has gained hundreds of miles … much of which is thick, multi-year ice."

In 2014, Cairo saw its first snow in 100 years.

In 2015, Oregon, like several other states, reached its coldest temperature in 40 years and Chicago saw the coldest days ever recorded.

In 2015, Antarctica reached the coldest temperature ever recorded anywhere on earth and in 2014 an environmental research ship was stuck in the Antarctic port that had always been liquid up until that time.

Besides all this, there is something else that EVERY LAST CLIMITOLOGIC knows.

100 Thousand Year Cycle

We know that there are these 100 thousand year temperature cycles that seems to be self-generated by the Earth. Every 100 thousand years or so, the Earth gets terrible cold, which sometimes shifts the rotational axis and causes extinctions and or problems. These changes and extinctions are done outside the emission of hydrocarbons and greenhouse gas. The "green gas" issue is fairly insignificant---especially CO_2 as there are very few solar wavelengths that are absorbed by this special and needed food for all plants. That is not to say changing the absorption of solar energy doesn't affect our weather and possibly change the course of our future. We'll look at some of the data in just a minute, but let me get back to a timeline and continue away from the dinosaurs and bring you up to date quickly so we can get a feeling for the Earth and the trauma it has endured and what you have been told about Venus bursting into its molten state from someone allowing Freon from their air conditioner escape on that planet. Venus did not have greenhouse gas runaway. What killed Venus was very different. The problem was no one could use Venus to hide things unless it was destroyed by runaway greenhouse gases.

They have done the same thing with the 100 thousand year thermal cycles on the earth rather than telling people the uncomfortable details of the Earth shifting on its axis fairly often; they made up some cock-a-mamy story that civilizations destroyed the ecosystems.

Earth Shift and Changing Thermal Patterns

As we think about the Earth thermal increases and decreases, we need to not only look at Cow and Dinosaur flatulence and Freon escaping; we need to look at the earth axis shift and what might have caused it. For this, we need to look at the Hawaiian Hotspot and understand that the trail of this important hotspot goes perpendicular to the rotation of the Earth.   As I showed before, 10 to 12 thousand years ago, the Earth shifted 30 degrees or so to where it is today.

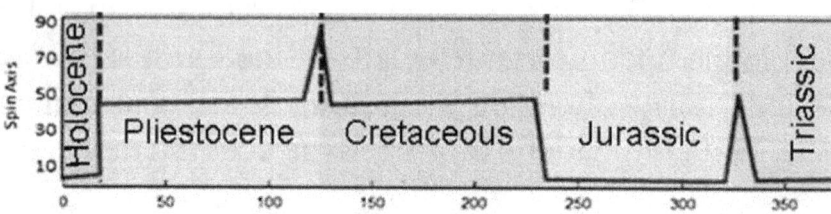

Scientists call this the Pleistocene Extinction and the atmosphere was in turmoil from something that happened on our nearest neighbor, Venus. I know you have heard this one. If you burn too many trees or use too much coal the earth will become like Venus with something we will call the Greenhous Effect--- greenhouse gases increase the heat which produces greenhouse gases, etc. etc. etc. Lie! Lie! Lie! The problem, like I mentioned before, is that there is a tiny bit of truth so it sounds right and good old Venus is sitting there to be a reminder. Once lush and tropical, it is now turned on its side, almost has its rotation stopped, a massive gash almost splitting it in two, and 700 degree temperatures that have boiled away the atmosphere, water, and anything else. <u>First let me say greenhouse gas cannot halt the rotation, turn it sideways or split the planet</u>. There was something else going on. That is not to say someone might have been on Venus before all this happened and used some underarm spray and got yelled at. I'm going to give you a super-fast overview of what happened up there.

Very Recent-First it was not extremely long ago [about 11 thousand years ago. Scientists have all wondered why all the artifacts they find are very young as the massive volcanic action that send molten lava all over the place and then they all stopped. We know it was very recent because none of the lava beds have been covered up, none of the river beds [including tiny tributaries] have been covered up, none of what looks like thousands of miles of roads are covered up, and massive winds in the upper atmosphere should have destroyed most of that evidence.

Equatorial Craters-Most of the major craters are along the Venusian equator so most likely something hit the surface that once revolved around it and was nearby. While this usually means a moon was around the planet there is no moon today.

Carolina Bays-Ten to 12 thousand years ago well over 500 thousands meteoric pieces peppered the eastern coast of the United States and the crater evidence is still extant. The craters are in a line as if it was a time when the Eastern Coast was the Equator. Very soon after the Carolina Bay meteors hit with horrible force, we know the earth axis shifted, the world became unsteady, water from massive tidal waves covered the world and almost complete extinction of life accompanied what we call the Pleistocene Extinction 10 thousand years ago.

Historical Reference-Dozens of ancient texts talk about the "planet with a wavy tail, or planet with a fiery tail, or the angry planet, or the fiery planet, or vain planet" that destroyed much of the land as if many saw the destruction of Venus during ancient times.

Biblical Planet Rahab- The Bible directly discusses the destruction of the Planet Rahab during an ancient war that saw 1/3 of the entire population of the world being killed. Soon, the entire planet was destroyed as Noah's famous surviving boat ride ended the Pleistocene extinction period. The ancient Biblical Texts describe a horrible War just before the End of the Pleistocene where 1/3 of the entire world population was killed.

Strafing Evidence-A number of in-line, exactly circular and same sized massive holes have been found on the surface of Venus. These could not be naturally occurring and appear to be bomb blasts.

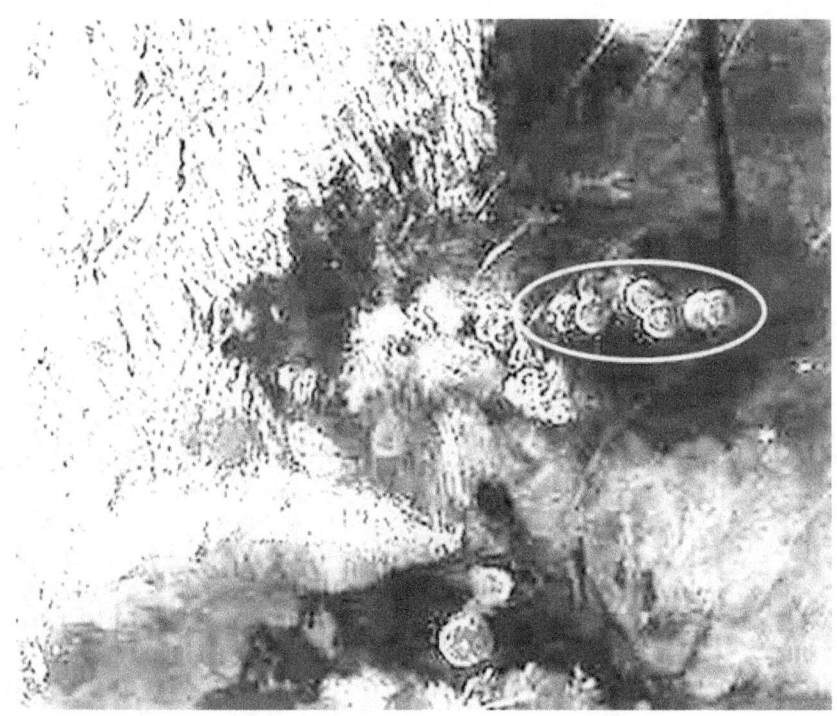

Later I will blow up the area shown and show how the strafing run had hit its target and possibly even destroyed a major roadway.

Pleistocene Extinction Was Not Thermal Runaway

I know you are thinking these are only coincidental, and I don't want to provide the massive amount of data that confirms these things here, but I do want to give you some brief data next. First let me give you a few of the Biblical texts describing the massive war and the destruction of the Planet Rahab/Venus. Over and over again it says the same thing. The Planet Rahab [Vain place] was completely destroyed. Huge pieces of debris were made as it shattered and the pieces became *"stones of fire"* or meteors.

Psalm 89:10 - *"Thou [God] hast <u>broken Rahab in pieces</u>, as one that is slain;"* [The pieces sound like meteoritic pieces. Especially as we read further.]

Isaiah 51:9- *"O arm of the LORD; awake, as in the ancient days, in the generations of old. Art thou not it that hath <u>split Rahab</u>, and wounded the dragon?"* [The Dragon most likely was the leader of some military group. Note the <u>idea that the planet was split</u> as is seen in the topographical map that follows.]

Job 26:12- *"The boastful Angel and his followers rebelled. Yahweh destroyed their dwelling places. He*

66

*divideth the sea with his power, and by his discretion <u>he</u> <u>smashed Rahab</u>. It was reduced to **stones of fire**."* [By this verse we could well believe that many people had made Venus their home before the disaster.]

Enoch 85 and Revelation 9-*"I beheld a single star fell from heaven-then I beheld many stars which descended and projected themselves from heaven to where the first star was."* [This could very well be the vision of many meteorites hitting the Earth.]

"Jasher" provides us with estimates of the destruction during the disaster. It indicated that 1/3 of the inhabitants of the entire Earth were destroyed. Here are the specific Biblical verses.

Jasher 2:5-6- *"-and the sons of men forsook the Lord all the days of Enosh [Adam's grandson] and his children; and the anger of the Lord was kindled on account of their works and abominations which they did in the Earth. And the Lord caused the waters of the river Gihon to overwhelm them, and he destroyed and consumed them, **and he destroyed the third part of the Earth,** and notwithstanding this, the sons of men did not turn from their evil ways--"* [This was a horrible war and Venus got involved. I know you are wondering how they got up there, but right now you look at the evidence.]

Isaiah 14:12- *How art thou fallen from heaven, O "Heylel" [**morning star**], son of the morning! How art thou cut down to the ground, which didst weaken the nations!* [Essentially, this is talking about parts of

67

Venus falling to the ground and weakening the nations. The idea of weakening nations sounds like something hurting the human inhabitants of the earth such as one would expect from the huge meteorite storm aftermath of an exploded moon of the nearby planet Venus. By the way, this Heylel term is used no place in the Bible except for this verse.]

Craters That Aren't Meteor Craters

The row of craters below left are not indicative of a meteor shower that would cause a random layout of variably sized blasts as the meteor exploded high in the atmosphere. These strikes are directed in a line on Venus. Here are seven blast areas in line. Someone was apparently trying to hit something during a strafing run of some kind. One thing that should be noted is that each of the blast areas is exactly the same size and perfectly round so the blasts could not have been random pieces of meteor unless each piece came from the same source, all happened at the same times and all pieces were the exactly the same size and density. This is indicative of bomb blasts where each bomb is the same number of megatons.

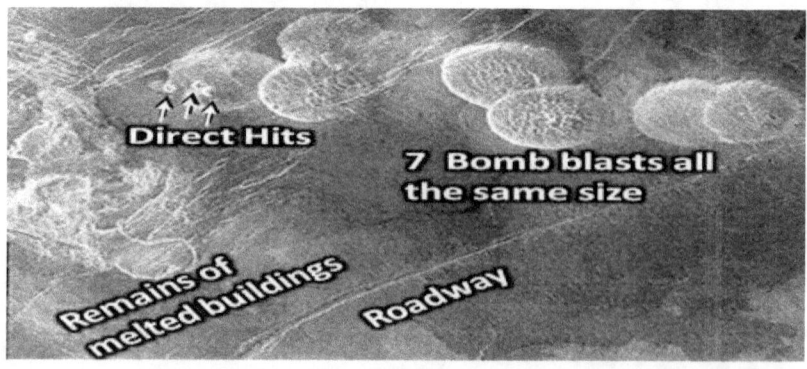

Images of Venus

Next, are a few of the many images showing how it used to be. The remains of massive waterways are still extant, but all the water was quickly removed when the Planet died. It was such a recent time that delicate tributaries, rolling hills, and majestic mountains [1.5 times as high as Mt. Everest] are still visible today. Also below is an image of the middle of Venus showing the massive split as the planet almost was cut in two.

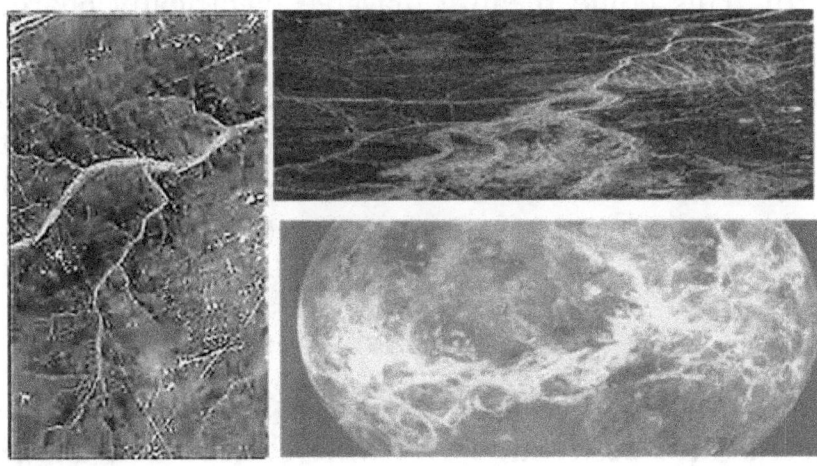

Argon-A main curiosity found by the Magellan probe was that the atmosphere of Venus contains high levels of the isotopes of argon, neon and noble gases. These high concentrations of noble gases could only mean that the current atmosphere of Venus is extremely young [on the order of thousands of years---say 10 thousand] because noble gases don't combine with other materials and escape easily into space; even with a thick atmosphere.

Plasma-To add craziness, in mid-1997, the Soho satellite detected a plasma structure issuing from Venus and it is long enough and in the right direction to almost reach the surface of Earth. The report described the structure as "stringy." Now for the interesting part; such a structure could only remain intact if a current were continuously flowing from Venus to the surrounding space via the plasma tail. "The initiator could very well have been uneven electrical charges between Venus and Earth. This could mean a massive electrically charged plasma burst reached Venus as our planets came too close together. I know that sounds science fiction, but Plasmas are used every day in our Fluorescent Lights, so don't discount them and Planets spinning all over the place WILL build up a charge and the chances of 2 planets having the same charge is almost impossible.

One Third of all Men Died in War

Even before the flood that ended the Pleistocene, there was massive destruction everywhere. Let's take another peek at Jasher to fill us in on some of it.

Jasher 2:4-7- and the sons of men forsook the Lord all the days of Enosh and his children; and the anger of the Lord was kindled on account of their works and abominations which they did in the earth. And [later] the Lord caused the waters of the river Gihon to overwhelm them, and he destroyed and consumed them, and he destroyed the third part of the earth, And in those days there was neither sowing nor reaping in the *earth; and there was <u>no food</u> for the sons of men and <u>the famine was very great</u> in those days.*

This is describing the horrors of war as those not being killed during a war are killed from famine and other devastation before the end of the Pleistocene.

__Enoch 106:13-16__. And I, Enoch, answered and said <u>in the generation of my father Jared the Anak people transgressed the word of the Lord</u>. And behold they commit sin and transgress the law, and have united themselves with women <u>produce on the earth giants</u> and the <u>earth shall be cleansed from all impurity</u>. Yea, <u>there shall come a great destruction over the whole earth, then</u> there shall be a deluge and a great destruction for one year. The __<u>temples burned with fire</u>__, And they shall remove them __<u>from the whole earth</u>__,

Notice in this section of the book of Enoch that there is a great destruction just before the great deluge. This is referencing the war and Meteor Storm that will lay hundreds of thousands of craters on the East Coast of the United States. It was the nature of the Earth thermal cycling that changed the temperature and the shifting of the earth that made it unlivable for a time----not gasoline engines.

Please forget I just told you there were people on Venus, that massive nuclear wars were fought during the Pleistocene, and that Dinosaurs lived a little over 100 thousand years ago. When you read about dozens of radio-active, unfossilized dinosaur remains being found today, TURN off your television.

Fake Fear Mounts

If people had anything to do with the global temperature variations, we would have noticed the change during this last Great War and all the rest that nobody told you about. With that, let's look at what you ARE being told with a little more detail!!!!!!! For one thing, they do not tell you about the details of these ancient conflicts that certainly pushed substantial greenhouse gas into the delicate atmosphere. Another thing that is not indicated is that there were so many people living in the world during these times, people COULD affect the atmosphere. Finally, what the fear mongers did see, they have misinterpreted and caused those who go against them great hardship. I'm not saying all those believing in the near term catastrophe of global warming by mankind are all deceitful, power hungry, people trying to introduce an artificial need for some special interest industry that they have massive interest in. What I am saying is _always follow the money_. If coal, oil, and gasoline are being targeted; what industry and what people might gain power or money?

Altruism!

I'm all for altruism and even directed efforts to truly save our planet, but there do not seem to be many of the leaders in this calamity who are sincere. It truly is a shame! New industries for Nuclear Power, Wind power Czars, and Solar Power Generals, were competing against the Oil Barons and Coal State livelihoods to support comfortable life styles around the world. Governments were pressured into providing "special incentives" for "green" industries in a race to halt the deadly "Global Warming" menace that was being initiated and continued overtly while feeding so many and so much bunk that large numbers of frightened people screamed for more regulation, more destruction of our coal and oil resources, and prison for all who would get in the way of saving our planet from the doom of carbon dioxide.

Nuclear power seemed to be well placed to be the leader, after the huge cost of putting in plants, the running cost was even less than that of coal, but problems soon arose, as accidents around the world put another fear into people and the inability of anyone to find a place for the "dirty water" effluent allowed more inefficient energy producing components to surge ahead. Please notice Geo-thermal and Gas are about 2 times as expensive as coal while wind is 3 times and solar is 5 times as expensive. I know you were NOT told this when you got your solar water heater, but subsidies are trying to artificially lower costs to get people used to them.

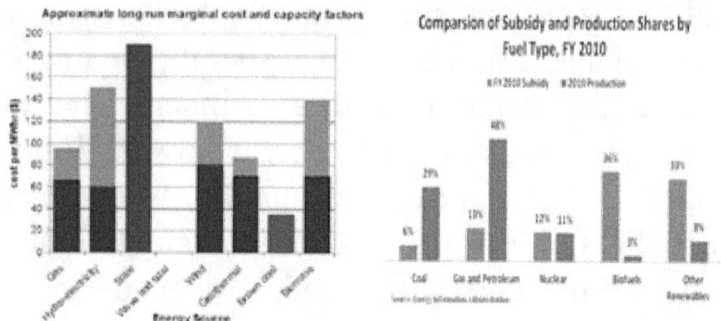

As shown to the left huge susidies are being provided for everything besides Coal, Gas, and nuclear energy with almost no useable production as the inefficiency cannot be eaisly overcome.

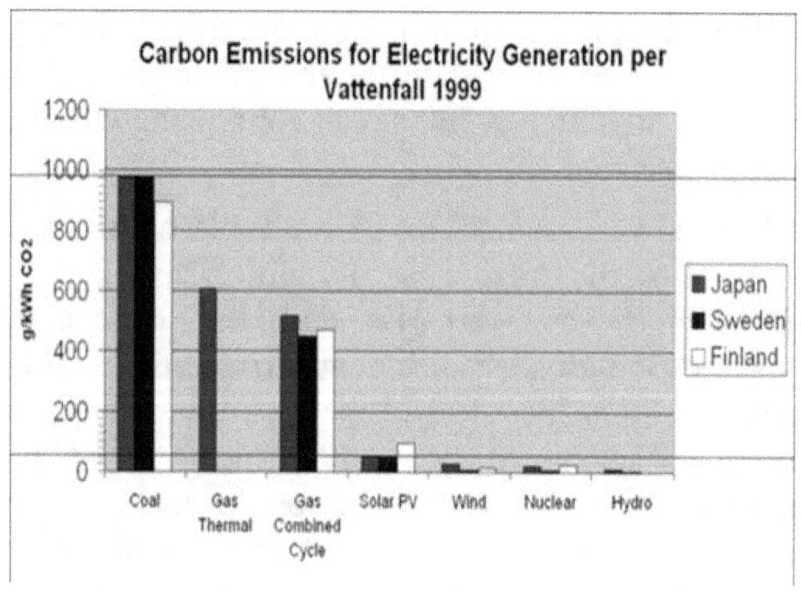

The preceding chart shows almost <u>all electricity is still made from coal </u>or a combination of coal and gas. If all the hydro-electric, Nuclear, Solar, Wind and all other electricity producing methods were destroyed, we would see almost no effect [except for fall-out from the nuclear plants]. Therefore; we need to be somewhat sure that eliminating coal will save us. After all; until 1990, we produced more coal than Asia, Europe, Soviet Union, Australia, South America, or Africa. Today, we are still the 2nd largest producer, but China has taken a huge step in energy independence as it increased coal production by 500% since 1990.

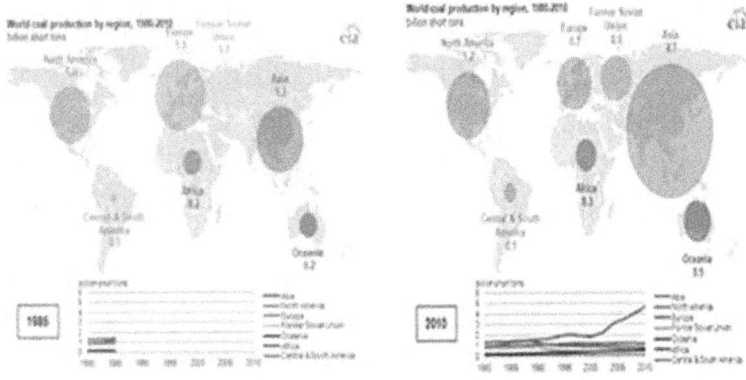

I know everyone is talking about how China is seeing how dangerous coal can be, but what they are worried about has nothing to do with global warming. If you burn too much in an area, CO_2 replaces breathable air. This is a completely different issue and not part of this discussion. Carbon Dioxide increases have been noted for some time, but tying it to global warming rather than looking a chem-trails and NAARP is a travesty. We will discuss the vilification of Coal and CO_2 as the greenhouse knock our chemical, but what about wind?

Wind Power Fiasco

As shown below, Wind Power capacity in the United States has increased by leaps and bounds or 1900% from 2000 until 2011. One would think that wind power was the savior of the planet. As shown previously, almost all this energy is so greatly subsidized, it is crazy to continue, but someone is making big money---- saving the world. Let's make sure he is really saving it.

The chart to the right shows that even with this enormous capacity growth, there is very little energy produced as little old Denmark produces 4100 percent more energy this way than the massively subsidized United States.

Avoided CO2 Emissions from Wind Energy

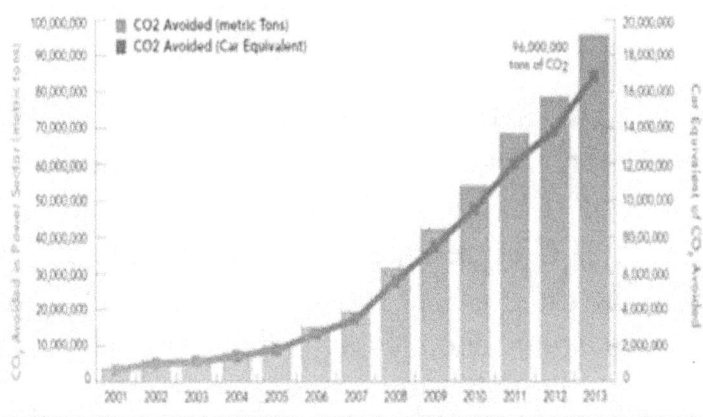

- In 2013, wind generation avoided an estimated 95.6 million metric tons of carbon dioxide (CO2)— the equivalent of reducing power-sector CO2 emissions by 4.4%, or taking over 16.9 million cars off the road.
- The 12,000 MW of wind power capacity under construction at the end of 2013 would reduce another 20 million metric tons of carbon dioxide (CO2) when it is operational — the equivalent of reducing power sector CO2 emissions by another 1%.

I thought it would be neat to show how these guys perk interest. According to this AWEA [Wind Industry] report. The world has saved 96 million tons of CO_2 getting into our air since 2001. Ha Ha Ha ---I laugh! Wind is doing nothing at all to make carbon dioxide go away and we can be glad it does not go away too much or we will lose our plants. Now we are fining that the continuous turbine noise is killing whales, seals, endangered birds and Bats at alarming rates and getting people sick as well. Let's see if Solar industries are also doing their part and cashing in.

Solar Power Fiasco

While wind is a travesty, solar power generation has been an utter failure as the following graphs shown. Soon it was the biggest cash cow for all sorts of unscrupulous Congressmen, Business Executives, and Scientists in the NOAA and similar organizations. The first shows solar energy subsidies cost <u>150 times</u> as much as the laughable wind power pursuits and **<u>121,000 percent</u>** more per megawatt than coal. The second one is even worse. It shows that our government is spending <u>200 percent</u> of what the wind power guys get to <u>0.007 times as much energy</u> as coming from the already inefficient wind power.

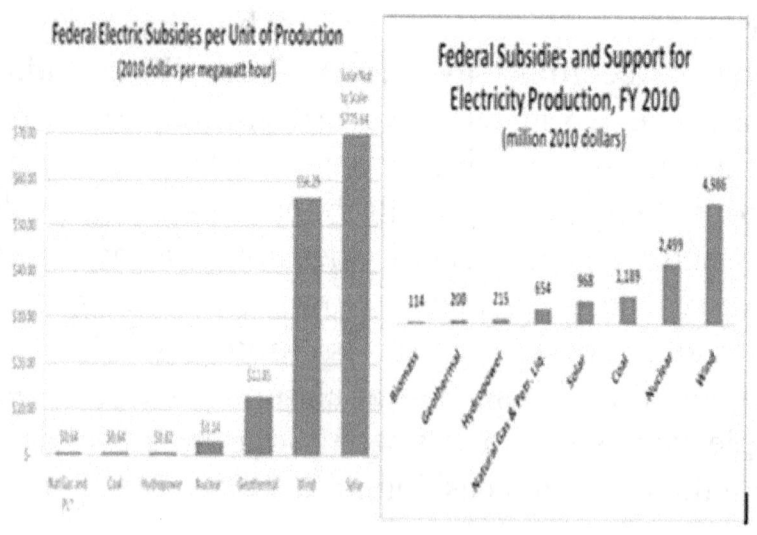

Unfortunately for you and me the fear mongers have told everyone that CO_2 was killing the Earth. Pressure from those being misdirected allowed awfull legislation to allow many to gain huge wealth art the expense of taxpayers. All they had to do was to fabricate a horrible calamity and we are paying the awful price. To make this worse, many of the companies gaining these subsidies are foreign with the highest concentration from China.

NOAA Deceptions

Luckily, we have at least 2 agencies looking out for us called NOAA [National Climate Data Center and National Oceanographic and Atmospheric Administration] and another one called the IPCC [Intergovernmental Panel on Climate Change] run by the UN. Sorry that was a joke. There is no question that these people are NOT. What we thought was 100 of the Top Global Thinkers [according to the *Bulletin of the American Meteorological Society*] we now know would do ANYTHING to advance their AGENDA no matter what the evidence shows. In no way are they worthy of ANY respect.

First Indication- ARGOS Dishonor

Scientists designed 3000 ARGO buoys that just floated around and took temperature measurements since 2003. These buoys show absolutely NO thermal increase [but the published data from NOAA somehow showed massive changes that the buoys "SOMEHOW" missed. The NOAA team decided that the information from them should not be used. One reason noted is that they weren't floating near the Arctic. Some might wonder why NOAA would have helped place these things and later decide they were stupid.

Some begin to suspect that there was a large network of politicians, corporations, and scientists that were conspiring to promote the fear of "global warming" . . . <u>despite evidence clearly stating no such "global warming" exists</u>. With only $22 billion being pushed into the global warming epidemic, you might wonder why some would try such a scare tactic.

NOAA Published Mistake Dishonor

The National Climate Data Center and National Oceanographic and Atmospheric Administration [NOAA] put out a chart showing there was no significant temperature rise in the United States from 1940 until 2010 and that the spring was the coldest in the 115 year record, but at the same time they told everyone to ignore this data and focus on eliminating the Coal Industry.

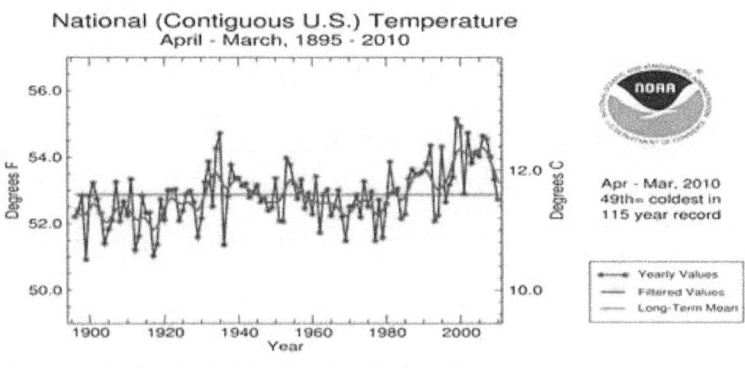

National Climatic Data Center / NESDIS / NOAA

After realizing their mistake they put out a completely different chart so that people would fund their pet projects associated with Earth annialation. By focusing on the slight rise since about 1960, the danger begins to look real. We are going to look into this chart that has been the foundation for many, many others and is still being used even after the fraud was fully exposed last year. You will be seeing this same massive thermanl ramp again and again, but I will show you the true slope for the original data from the Bouys and sattelites.

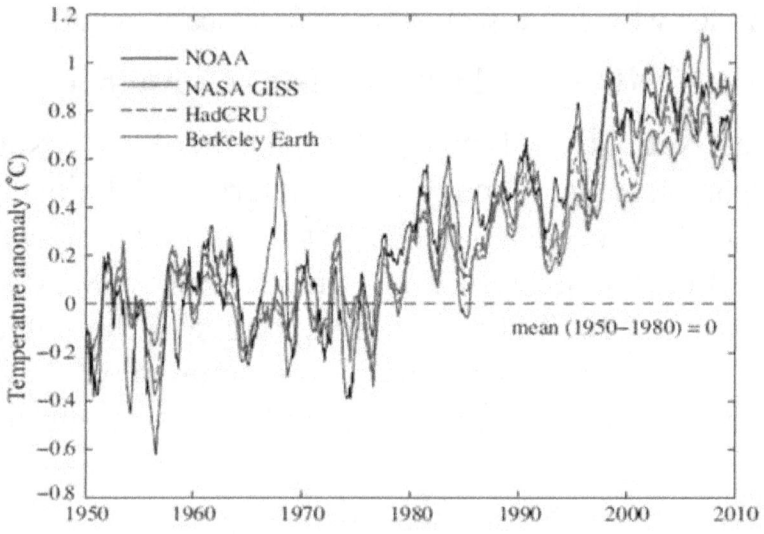

17 Year Cooling Detail Dishonor

NOAA got fancy with this one. According to NASA's own data, the world has warmed 0.36 degrees Fahrenheit over the last 35 years, starting at the fairly cold year-1979. Even this would show a massive increase of 0.1 degree per year over that short time. The NASA Remote Sensing Systems data also shows that since 1998, the average temperatures around the world have been steadily decreasing as shown below.

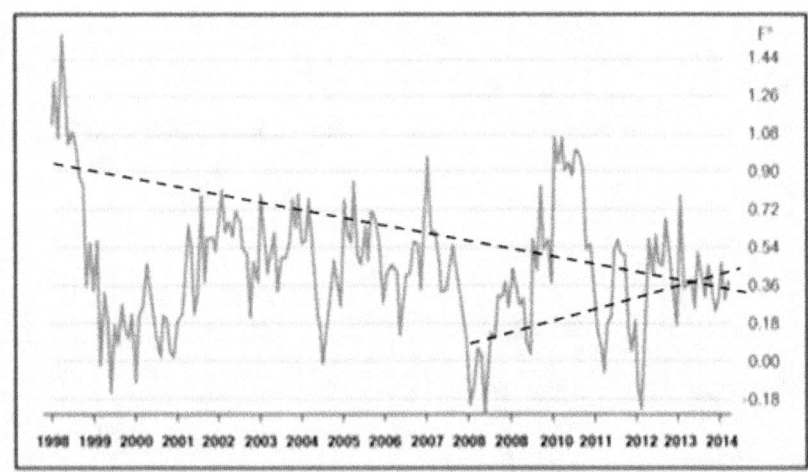

According to this graph, the world is 1.08 degrees cooler than it was in 1998. NOAA has not used this data when "Informing the World" about the condition of our Earth. To make it look like horror, just truncate until a rise is shown [2008 until today shows a 0.32 degree rise]. This is not the new data I mentioned to correct the previous longer term graph. That is next.

Just Plain Lying Dishonor

NOAAs current US graph is shown below left [same as before]. *Now we know it is all a lie.* Note that there is a discontinuity at 1998, which doesn't look right. Globally, temperatures plummeted in 1999-2000, but they didn't in the US graph. Note that measured data below right shows that by 2008, temperatures were back down to the 1989 level. But in the NCDC data, 2008 is half a degree warmer than 1989 making the temperature LOOK like a disaster when there is almost no change at all. Please note that the faked chart to the left has been used to justify an enormous number of "charts showing the destruction of our world.

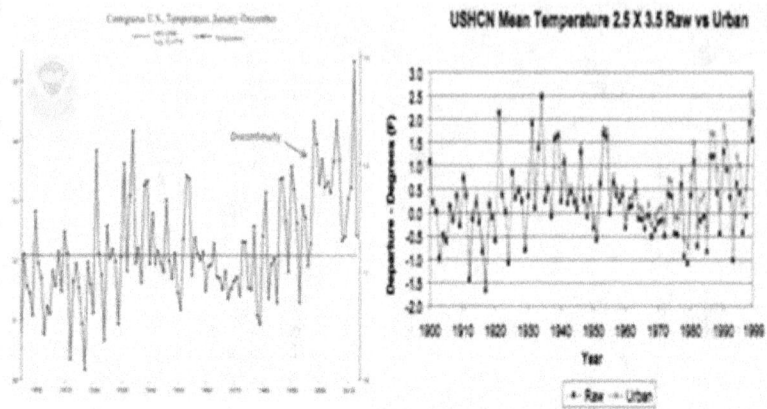

By putting the 2 together we can easily see the treachery. The top graph is from RAW data and the bottom one is the "doctored" chart making everyone want to give money to Green Technologies to protect them from this FAKE temperature rise.

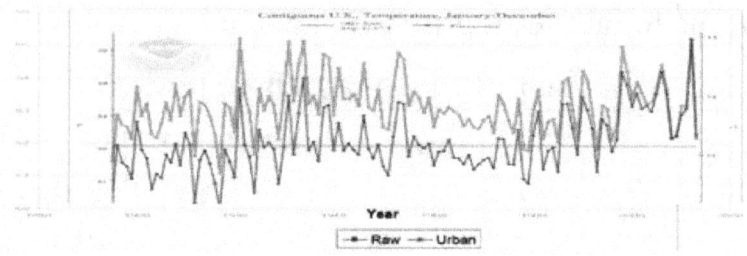

Bottom line is that NOAAs US temperature record is completely broken, and meaningless.

Adjustments that used to go flat after 1990 now go up exponentially. Adjustments which are documented as positive are implemented as negative to amplify fear for monetary gain.

1940 Spike Removed Dishonor

The best way to describe the subterfuge is to talk about the 1940s heat wave. This was a huge thermal "spike" occurring as the Earth recovered from the mini-Ice Age. If you were somewhat devious, you would like that spike to go away as it shows the temperature today is not significantly different than 1940 levels. This can be done 2 ways.

1. Show the thermal rise since 1965. This is the favorite one.

2. Change the 1940 peaks to smooth out "anomalies that don't go along with the "THREAT". In this case it seems unthinkable, but this is another way to make people buy Solar Cells for their homes. [Especially when the US Government pays for most of the installation to eliminate the use of Coal.]

We now know "cooling the past" adjustments have been carried out in the Arctic region to an unbelievable level of sabotage. Nearly every current station from Greenland, in the west, to the heart of Siberia has been altered in this way. The effect has been to remove a large part of the 1940's spike. I think the chart below left shows you how it helps breed fear. The actual data has high temperature bumps centered in 1940 and 1960, but the common practice is to eliminate the annoying bumps to make the temperature seem to be ----OUT OF CONTROL. Sometimes a little of the spike data is kept in but the density is reduced to show an average increasing slope is more terrifying as shown below right.

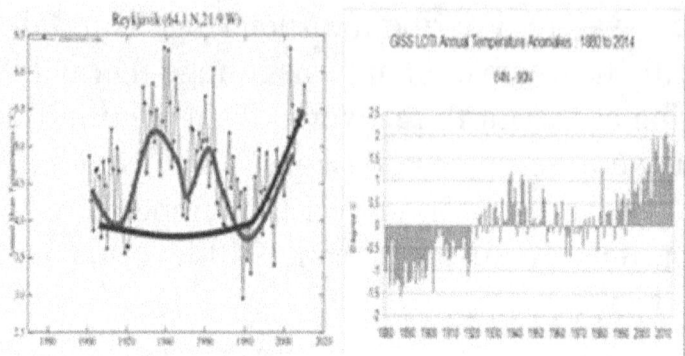

After the practice was revealed from some cleaver undercoverwork and retreval of internal EMAIL traffic, we have started to see a few of the practicioners coming clean about their part in the underhanded fear mongering.

Ice Core Manipulation Dishonor

This one I mentioned before. Adding aerosol CO_2 to Ice Core data to make global warming look bad ---after just publishing details indicating it takes hundreds or even thousands of years for Aeresol CO_2 to finally get absorbed by the ground.

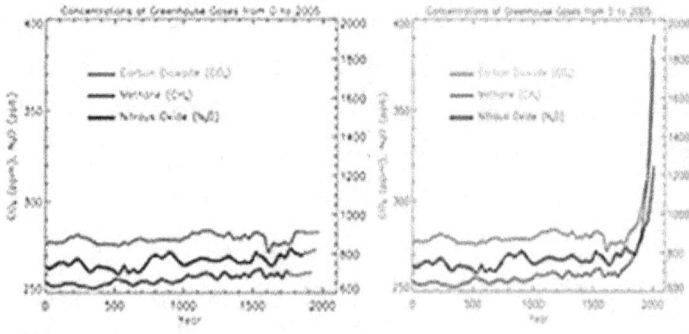

The first chart shows the Ice Core data to about 1980 and the second BOGUS one shows data retrieved ffrom Mona Loa Hawaii Aerosol CO_2 tests which have nothing to do with Ice Core data.

No Ocean Rise Dishonor

Even last year they PUBLISHED huge increases in oceanographic water height caused by this global warming monster and then this year they had to publish this bogus, dirt must be a sponge that says 440% of what would have drowned countries was soaked into the dirt making the oceans stay low. We may never know if these people are just stupid or are trying to protect some secret while cashing in on the popularity of global warming "research", but there is no question that they have done a huge disservice to all mankind---- and I really mean everyone around the world. Besides these first things, let's review a tiny portion of the NOAA subterfuge.

Scandalous Actions

Called *Mike's Nature Trick*, perpetrators of the change in temperature charts that greatly fed the fire on global warming fear are now known and the act has been proven. It seems none went to prison or were even fined for what is perhaps the worst type of criminal extortion and conspiracy. While generally these guys are protected by those giving unbelievable wealth, the sad fact is there is no specific law for holding the world hostage to fear and with that fear being able to extort BILLIONS of dollars from the innocent.

Coming Clean

Dr. Phil Jones — a leading "global warming" advocate at the United Nations — admitted that he used *"Mike's Nature trick" in a 1999 graph to "hide the decline" in temperature.* He was never charged with any crime---nor was the originator of the deception.

Dr. Stephen Goddard had this to say: *"The National Oceanic and Atmospheric Administration (NOAA) has been "adjusting" its record by replacing real temperatures with data "fabricated" by computer models."*

Dr. Robert Stavins, who helped write the 2014 United Nations Climate Report, revealed that *"politicians demanded he change and edit parts of the report to fit their needs!"* You would think his changes would land him in prison, but he is still one of the "respected" scientists.

Dr. Kevin Trenberth — One of the NOAA elite finally admitted, *"The fact is that we can't account for the lack of warming at the moment and it is a travesty we can't."* If there is no warming all of the billions of dollars going to the select would dry up and the United States could spend money helping the majority of Americans instead.

D. James Lovelock, was awarded the Wollaston Medal for his climate work in something he called Gaia Earth. He was one of the first to proclaim that *humanity would soon end due to global warming from CO2.* After the last 15 years with no global warming, he has admitted

to being "alarmist" about climate change and says-
"Other environmental commentators, such as Al Gore, were too...The problem is we don't know what the climate is doing. We thought we knew 20 years ago. That led to some alarmist books – mine included – because it looked clear-cut, but it hasn't happened."

Al Gore Gets Rich

Al Gore has done a tremendous disservice to humanity and gained millions of dollar doing it. I'm talking about the 2007 Nobel Peace Prize winner who gained fame in his push to place fear in the hearts of all concerning the Faked" global warming scare we help push.

His "Ice" Lie

This guy stated the following, *"The North Polar ice cap is falling off a cliff. It could be completely gone in summer in as little as seven years. Seven years from now [2007]."*

"Man Caused CO$_2$ Killer" Lie

Al Gore claimed *CO$_2$ emissions from Human factories were destroying our world as massive amounts of factory effluent were making the Earth's temperature go out of control and 97% of scientists agree it's real, it's man-made, and it's dangerous.* Certainly, he knew the satellite data and all the rest, but he was making a fortune. He also knew many scientists were begging for people to listen to them as Mr. Gore misrepresented everything.

I know it seems like everyone you meet indicates that man-made green gases are killing the earth and there are charts showing that most scientist agree and thanks to a fabulous campaign of deceit, the "normal population" now believes we are destroying our planet, but Please just read the book and then make up your mind rather than blindly listening to some reporter on CNN or the president of some Solar Cell manufacturer tell you how much less Carbon is being introduced by giving up underarm spray or taking a bicycle to work or whatever. Have you ever heard the term "follow the money"?

Give Me Money

Al Gore, was worth $2 million when he left the Vice Presidency in 2001 and now, after investing almost entirely in a small number of "GREEN-TECHNOLOGIES" raping governments out of fear, Gore is worth close to a $billion. The massive artificial funding of these horrible industries goes away if global warming is not pumped up, so Gore seems to do whatever is needed to make his "investments" thrive when there is no way for them to thrive. Lying, misdirecting, lying some more; it's a business of fear and the art of misdirection for money.

Scandal After Scandal

While there are many "me-toos" associated with Al Gore, let me talk about one we know fairly well. His name is Barrack Hussein Obama and he has done more to cripple our natural resource independence of any President in my opinion. While never getting a Nobel Peace Prize for Global Warming, he certainly has been doing his best to excite our population. He tweeted last year the following in support of Al Gore's statistic: *"97% of scientists agree: climate change is real, <u>man-made and dangerous</u>."* Thank you Mr. President. Let's just destroy the coal industry right now just like you promised the Europeans you would do. Possibly this was to cover up testing by the HAARP group that massed up the glaciers in Alaska or maybe he is buying into the craziness. Luckily, the Wall Street Journal reported the following, *"The assertion that 97% of scientists believe that climate change <u>is a man-made, urgent problem is</u> <u>a fiction</u>."* When further review was done, it was discovered that many did believe the earth was getting warmer, but a mere 1% of scientists believe <u>human activity is causing most of the climate</u> change. Unfortunately, that was in 2006. Today that has changed as most seem to have been brainwashed into believing this horror.

91

Not to fear, a petition was signed by more than 31,000 "regular" scientists that states *There is no convincing scientific evidence that human release of… carbon dioxide, methane, or other greenhouse gases is causing or will, in the foreseeable future, cause catastrophic heating of the Earth's atmosphere and disruption of the Earth's climate."*

Diversion from HAARP

The popularity of fear has worked its way into every area of our society and President Obama increased the fear by saying, *"We want our children to live in an America that isn't threatened by the destructive power of a warming planet."* --- All the while he sanctioned some of the worst experiments to our survival that are only recently being reduced and as the experiments are reduced; the Arctic Circle is starting to regain its once cold glory. Unfortunately the HAARP experimentation has not completely halted as I will show in a little bit.

It is so criminal what has happened. I know my little book isn't going to help much, but hopefully some will begin to see more of the lies and stop some of this madness as we try our best to bankrupt our country on crazy scheme after crazy scheme. Money is obtained by many ways but the most insidious is by something called loan guarantees.

DoE Diverts Suspision from HAARP

The current Department of Energy failed Loan Guarentee Program is split into 3 types of loans as shown in the graphic below, but all require the public fear of global warming to keep the failed industries from having any chance at paying some of the money back and not forcing a ressession era.

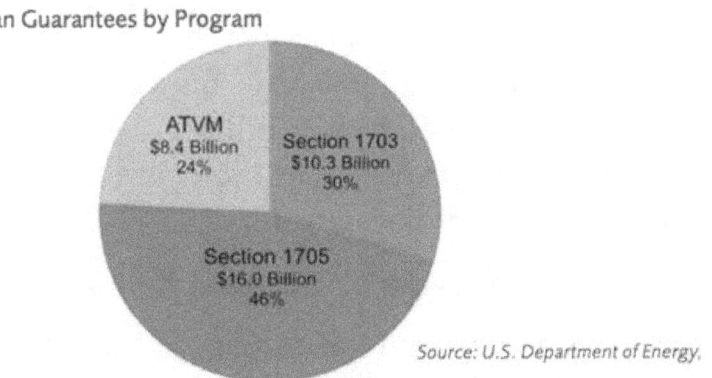

Loan Guarantees by Program

ATVM
$8.4 Billion
24%

Section 1703
$10.3 Billion
30%

Section 1705
$16.0 Billion
46%

Source: U.S. Department of Energy,

Advanced Technology Vehicles Manufacturing [ATVM] dollars were secured fearing gasoline vehicles would turn our planet into a deathtrap. Recovery and Reinvestament Act Section 1705 was to reduce effluent from bad industries that would make our planet die and build substitute capabilities. Section 1703 seems to just be a way to prop up Solar Cell manufacturing no matter what the cost. Forget scandal, this mess has turned into calamity. The visible dollars are shown as 34.7 Billion, but the total cost is an extremely difficult to estimate and consumers are repaid to put in solar, geo-thermal, wind and other components to prop up these fear built industries. Section 1703 was almost totally for solar power generation with one company NRG Energy receiving almost $4 Billion alone, with Abengoa receiving about $3 billion.

These don't include the Department of Treasury grants under something called <u>1603 Grants</u> which <u>went to the same companies and other grants under "Recovery and Reinvestament Act funds [Section 1705]</u>. Quoted in the *New York Times* recently, NRG's chief executive, David W. Crane, explained how his company and its partners have secured <u>$5.2 billion</u> in federal loan guarantees, plus <u>hundreds of millions in other subsidies</u> for four large solar projects.

Obama's "Solyndra" Scandal *[Solar Cells]*

Obama's Solyndra green energy initiative cost taxpayers $500 million. He, in essence, *used taxpayer money to finance his re-election campaign by funneling it through Solyndra.* He received $1.25 million in campaign contributions from Goldman Sachs and George Kaiser who took over Solyndra when it failed, so our President insured $500M would be returned for that campaign donation. Solyndra not only gained the $500 million 1705 fund guarantees but also multiple millions from the Ex-IM Bank to help sell the company.

DoE's "First Solar" Scandal *[More Solar Power]*

Here is a funny one. We gave "First Solar" multiple millions in "loan Guarantees then we loaned one of their subsidiaries $192.9 million to buy First Solar panels from themselves. That wasn't enough so they were given a $16.3 million loan in 2010 and another $646 million in 2011. Then they got another $547.7million loan from Ex-IM bank to subsidize panels outside the US. Somehow they received $4.5

Billion in government loan guarantees, and were able to give almost $200,000 to Democratic campaigns. Where did they get the contribution money?

"Abound Solar" Scandal *[Solar Cells]*

Another scandalous pay-off was to Abound Solar. Obama got a federal loan of $400 million for Abound Solar, in 2010. Just like the others, it went bankrupt after Patricia Stryker got the money she "needed". She contributed *$500,000* to the *Coalition for Progress* and another $85,000 towards Obama's inaugural committee as her bankruptcy payout was huge.

"GE" Scandal *[Wind Turbines]*

GE makes wind turbines and receives tens of Millions in what are called "Green Energy Credits". Because they received all these "credits" and saved our world, they didn't have to pay taxes at all in 2011.

"A123 Systems" Scandal *[Lithium Batteries]*

A123 Systems, a nanophosphate Lithium Battery source supplier, gave their lobbies $1M to get $280M Federal Assistance. Some might wonder if they only needed assistance to pay for the lobbyist???

"Johnson Control" Scandal *[More Batteries]*

Johnson Controls **"won"** a $299 million stimulus grant to build two battery plants, but, instead, it is running only one plant at ½ capacity and pocketing the rest of the money.

"Fisker Automotive" Scandal *[Electric Cars]*

Fisker Automotive was having the usual "can't make a profit and still keep all those giving me money happy blues so the DoE lent them <u>$529 million</u>. Lucky for us, it got so bad in 2011 the DoE halted the money transfer at $193 million.

Harry Reid and "Ormat" Scandal *[Geo-Thermal]*

Retiring in 2015, one scandal that Harry Reid is running from was reported on recently in both the *Washington Free Beacon* and *Courthouse News*. Ormat Technologies owns and manages geothermal plants in California and Hawaii. Reid and various others got them <u>$136 million</u> in economic stimulus funding. The executives essentially stole the money and didn't even present new green technologies at all. As reported in the Free Beacon, *"Reid bragged about securing Ormat another <u>$350 million</u> loan guarantee from the Department of Energy"*. For some reason, Ormat executives have generously supported Reid with donations for his election campaigns and causes.

$22 Million Or Is It More?

We could go on and on and on and on. The fear generated by the NOAA criminally imaginative "scientists" with their creative redistribution of temperatures to make our world seem hotter has gone a long way in reducing money we could have spent on industries with true potential. There is a figure typically associated with Green Energy distributions to scientists, agencies, and Stock Owners, but $22 billion is just what is *spent* on these "global warming" initiatives.

According to Forbes, the total cost to the citizens of the United States is <u>really $1.75 trillion annually</u>. The U.S. Energy Information Administration says- *the green Energy regulations alone could ultimately cause gasoline prices to rise 77% over baseline projections, send 3 million Americans to the welfare line, and reduce average household income by a whopping $4,000 each year.*

Not to be ignored, the Congressional Budget Office [CBO] estimated *the default on DoE nuclear loan program at 50%,* but guarantees the loans anyway. This allows banks to gain profit if the business does survive and the American public to have all the risk and loss.

Stop Crying! We can make up the wasted $1.75 Trillion other ways like not helping the Americans who lost their jobs or small business or the things than made America great.

You might be asking, "Why in the world would our government do all this?" The answer is "diversion". The science community is so bamboozled that they make detailed models with all the data they gather--- assuming CO_2 was causing the disaster, and, because they are not working on the correct data point---- they cannot make a model no matter how great they are in mathematics.

Modeling Disasters

Besides the outright lying, and giving taxpayer money to Green industries to make them seem reasonable; if you want to scare someone, use a model showing "Expected Disasters". I'm not talking about one or two. <u>My last count was that 73 of these things</u> were being touted as gospel. Don't get me wrong; these are important; the problem is they all failed and caused more alarm and inappropriate action as the forward vision was twisted and NOAA used these failed models to plan for a global warming disaster.

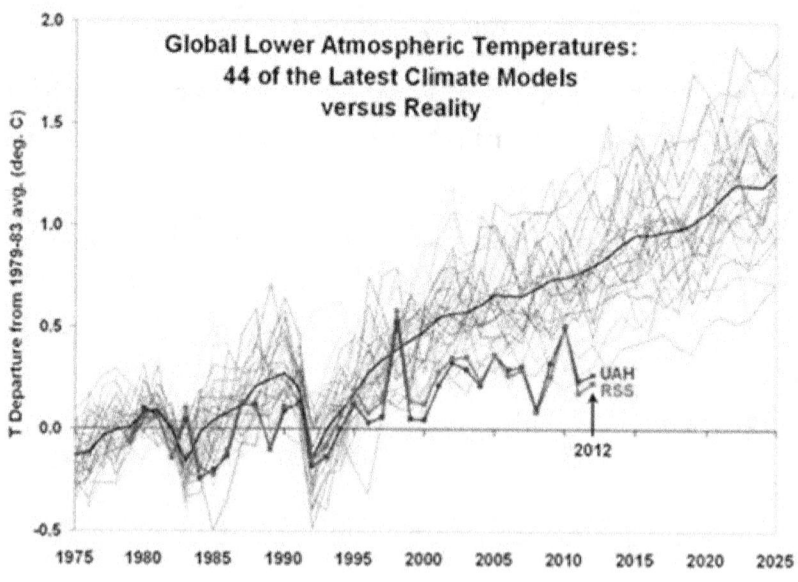

As shown above, 44 of these things are compared to actual Seattleite temperatures through 2012. The ONLY time the 'predicted' Mean World Temperatures were close to actual measured mean values was in 1979 – 1982. The actual temperature rise "estimated" from 2005 to 2012 was <u>250% higher</u>. None of the models is accurate. The trend as shown indicates that the predictions will get worse and will increase in inaccuracy. Clearly, the selected testing elements are not correct. Let me say this once more. The attempts are valiant and models are needed for prediction and research, but the scientists are not using the real thing that is upsetting our weather. For someone to blankly state that burning coal will kill us all with nothing to back it up is almost criminal. The actual measurements of mean temperature clearly show that temperature decreases with increasing elevation and these inappropriate and dangerous models do not. The frustration is well noted among those calling themselves Climate Scientists. Even James Hansen tells of his frustration about not understanding weather.

Robert Kaufman, climate scientist: "...released a modeling study suggesting that <u>the hiatus in warming</u> could be due entirely to El Niño and increased sulfates from China's coal burning." Can you imagine the Gaul??? The reason the models don't work is the Arctic warming has nothing to do with CO2.

Martin Wild, climate scientist: "During the 1980s and '90s, the rapid decline of air pollution in the United States and Europe dominated the world's aerosol trends. While those emissions have continued to decline in the West, returns, from a brightening standpoint, ... "It's not an obvious overall trend anymore,..."----It never was a trend---we simply made a model of the events without knowing the cause or trying to hide the cause. There was no significant decline in air pollution as just as many cars and jets were traveling all over the United States and Coal is what we have been making electricity from.

Susan Solomon*, climate scientist: ""What's really been exciting to me about this last 10-year period is that it has made people think about decadal variability much more carefully than they probably have before," ...Solomon had shown that between 2000 and 2009, the amount of water vapor in the stratosphere declined by about 10 percent. This decline, caused either by natural variability — perhaps related to El Niño — or as a negative]feedback to climate change, likely countered 25 percent of the warming that would have been caused by rising greenhouse gases..."* I don't know what she was smoking but this gobbly gook is simply mind boggling.

James Hansen*, climate scientist: "All the climate models, compared to the Argo data and a tracer study soon to be released by several NASA peers, exaggerate how efficiently the ocean mixes heat into its recesses....that climate models have been overestimating the amount of energy in the climate,... "Less efficient mixing, other things being equal, would mean that there is less warming 'in the* *pipeline,'" ....it also implies that the <u>negative aerosol</u> <u>forcing</u> is probably larger than most models assumed."* Just think about this stupid statement. He is saying if you used my model everything would be on track. The stupidity is he never made a model and it wouldn't have worked if he had as all seemed to try to ignore everything happening in Alaska, Sweden Russia and Puerto Rico. All 4 were heating the ionosphere according to what they did admit to and many other things that they identified only as experiments.

Judith Lean*, climate scientist: "The answer to the hiatus, according to Judith Lean, is all in the stars. Or rather, one star...<u>Climate models failed to reflect the sun's cyclical influence on the climate</u> and <u>"that has led to a sense that the sun isn't a player,"</u> Lean said. <u>"And that they have to absolutely prove that it's not a player."</u> According to Lean, <u>the combination of multiple La Niñas and the solar minimum, bottoming out for an unusually extended time in 2008 from its peak in 2001, are all that's needed to cancel out the increased warming from rising greenhouse gases."* Wow! This woman thinks CO2 that cannot absorb sunlight does more than the SUN. Maybe she has a smaller sun that I'm used to.

John Daniel*, climate scientist: <u>"We make a mistake, anytime the temperature goes up, you imply this is due to global warming,"</u> he said. "If you make a big deal about every time it goes up, it seems like <u>you should make a big deal about every time it goes down</u>."* Finally we have someone who at least makes sense.

No Model Can Work When the Cause is Hidden

While there is evidence that even those pushing human made global warming really have a difficult time finding facts that support that idea. They keep searching and some modify data, but most seem to be in love with the idea of cleaning the earth and protecting our environment by eliminating CO_2 that cannot harm anything. What can do immense harm is the sun. Additionally, too much water vapor, too much Oxygen and Ozone, and people messing with the Ionosphere can cause substantial issues. The modelers want to protect the Earth "from man" rather than protecting the Earth FOR man. They seem to have blinders on for apparent problems and fixated on the fact that we MUST be the reason for problems. Let's continue looking at their data interpretations.

Boiling Ocean "Mistake"

James Hansen, that climatologist I keep bringing up, was at it again. Look closely at the graph below this is what "boiling" oceans look like after some 1.3 trillion tons of CO_2 emissions poured into the atmosphere since 1850. The huge increase in CO_2 levels shown as a dotted line had absolutely NO effect on the Oceans that have stayed the same temperature. James Hansen's belief of CO_2 caused global warming is not supported by the tropic's data in the least nor is his crazy prediction of boiling oceans. You can keep driving you gasoline car and the earth will not even know it.

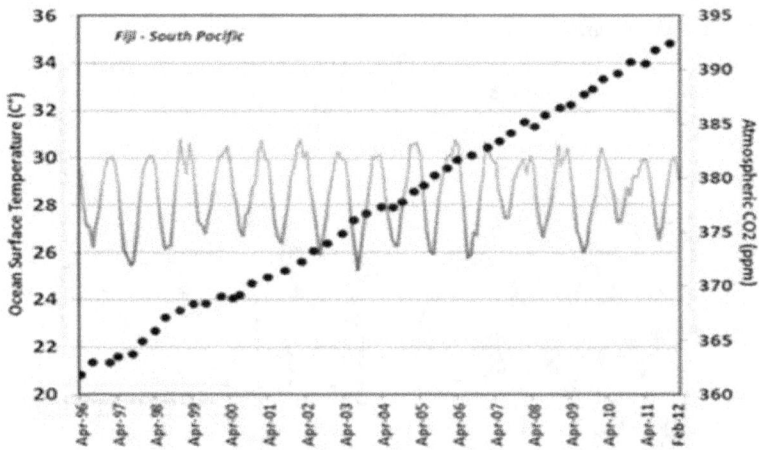

More Arctic Warming Detail

We are told that the biggest sign of "climate change" is the rapidly warming Arctic. It is even called the world's thermometer; proof that global warming cannot have stopped. Certainly, the evidence of this from the GISS satellite is persuasive for the Arctic, but not so convincing for Antarctica. The charts are of the average temperature between 1951 and 1980 compared to the average temperature in 2014. The Arctic is getting warmer just as the Antarctic is getting colder. Nobody is showing you the bottom of the Earth. Maybe we should look farther.

GISS Surface Temperature Analysis

Global Maps

Sources and parameters: GHCN_GISS_ERSST_1200km_Anom1212_2014_2014_1951_1980_POL

Annual D–N 2014   L–OTI(°C) Anomaly vs 1951–1980   0.66

3⁰ temperature Increase   3⁰ temperature Decrease

Substantial Arctic Change is NORMAL

It is well established that the Arctic warmed up rapidly during the **<u>1930's and 40's</u>**, before temperatures plunged in the <u>1960's and 70's</u>. James Hansen knew this as shown in the graph below from his 1987 paper *"Global Trends of Measured Surface Air Temperature"*. In the chart we see the Arctic temperature rise by 3^0 over a 60 year period before dropping again around 1940. If you remember, the article I provided from 1922, the same "using too much coal conspiracists were claiming the end of the world when they saw the rise.

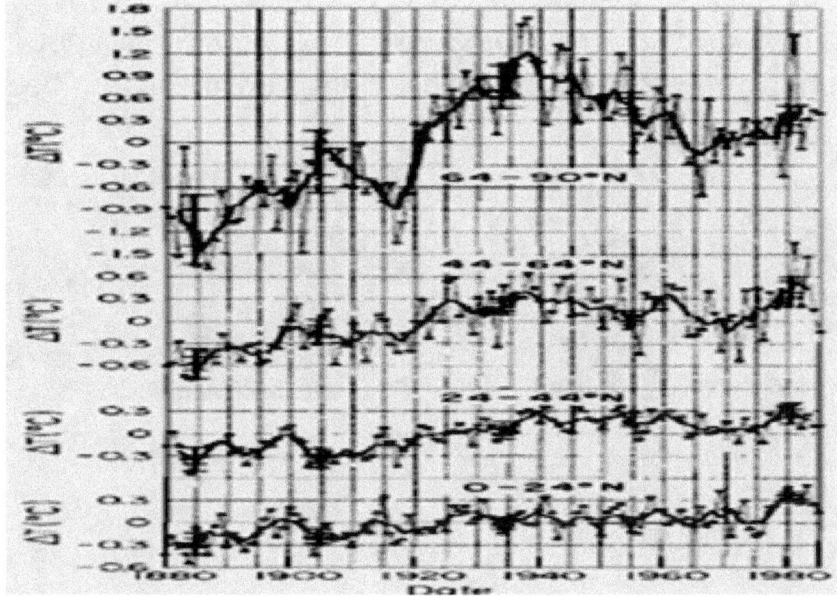

For the next 25 years the temperature dropped again by about 1.5^0 to what the alarmists use as the "NORMAL" Arctic temperature. Notice the temperature between 64 and 90 degrees latitude [Near the Arctic Circle]. Also notice the temperature near the Equator had a very slight dip in the Ice Age and a slight increase to 1980. While I think we need to be more responsible about the Ionosphere "experiments and cloud seeding, the current loss of Ice in the Arctic seems to be sort of a "correction" rather than a catastrophe.

Greenland Ice is Returning

In **Greenland**, Tavi Murray (Swansea University) reported that *two of the major glaciers, Helheim and Kangerdlugssuag, had slowed significantly by 2006, and reported that in 2007-08 that there has been a 'synchronous switch-off' of speed-up of the 14 largest outlet glaciers in southeast Greenland* ... not a 'runaway acceleration' that has erroneously been reported.

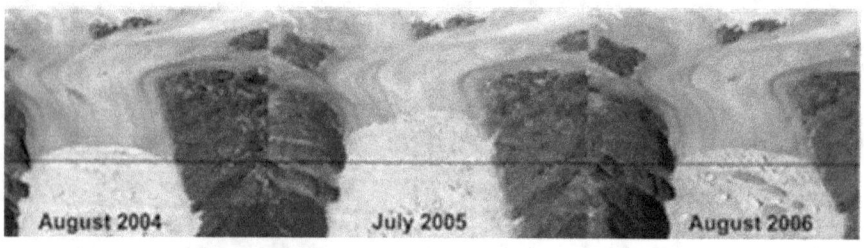

The graphic shows the Helheim Glacier's flow to the sea sped up in 2005, as evidenced by the 5-kilometer retreat of its leading edge, but by 2006 it had slowed back down showing the minor temperature had little effect even in the Northern hemisphere.

More Antarctic Detail

Most of the Antarctic <u>ICE is expanding, not melting.</u> This is contrary to widespread public belief. The *"Scientific Committee on Antarctic Research of Antarctic Treaty"* nations reported recently in Washington that, *'cooling at the South Pole was significant in recent decades'*. The eastern part of Antarctica is four times the size of west Antarctica, and <u>ice losses in west Antarctica</u> over the past 30 years have been <u>more than offset by increases in the eastern part</u> of the Ross Sea region. Ice core drilling in the fast ice off Australia's Davis Station in East Antarctica by the Antarctic Climate and Ecosystems Co-Operative Research Center shows that last year, *<u>the ice had a maximum thickness of over 6 feet is its densest in 10 years.</u>* The <u>*average thickness of the ice at Davis since the 1950s is about 5 feet*</u>.

Antarctic Sea Ice

It is reported that <u>Antarctic sea ice has *increased 4.7% since 1980*</u>. <u>Over 90% of the World's glaciers are in the Antarctic and they are growing.</u> The following image [left] shows the 2009 Ice extent compared with the average sea ice distance [Thin line]. This is 90% of the World's ice and 70% of the World's fresh water. Sea ice around Antarctica varies from about 8 million square miles in September or October to about 1 million square miles in January. The following image [right] is from NASA is of sea ice maximum image from the Nimbus 7 Scanning Multichannel Multiwave Radiometer (1978 to 1987) showing a similar increase. Yes; the area just below South America is showing a slight reduction in Ice, but the rest is getting more ice.

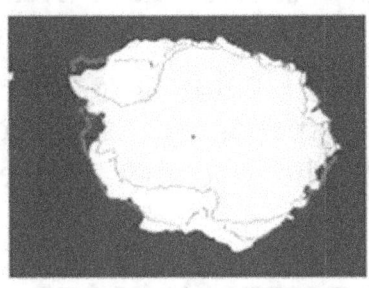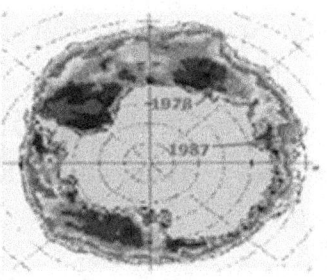

The next image [left] is a NASA composite map of Antarctica showing areas of greatest warming as the darker region. The rest is colder. The Wilkins Ice Shelf lies off the peninsula in the top left corner. Antarctic is approximately 14,000,000 square kilometers and about 8,000,000 square kilometers of shelf ice melts and reforms each year.

By picking the small area where the average temperature is getting slightly warmer for your FAKE GRAPH] it looks like Antarctica is losing ice, the oceans are getting deeper and the world is in trouble.

The thing to know is that the data [while it can be considered correct, is a total lie.] The graph to the right shows the slight increase in average Ice levels in the southern hemisphere. No one is showing us that graph.

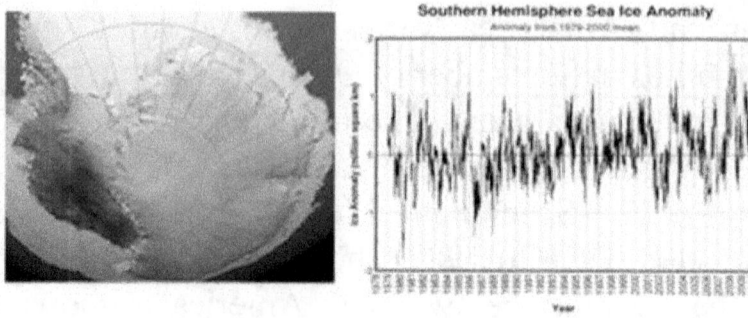

The next images not only shows the increase in intensity of the cold between 2014 and 2015, but also shows the <u>number of super cold days increased</u> significantly and 2015 experienced the coldest temperature ever around the October November timeframe.

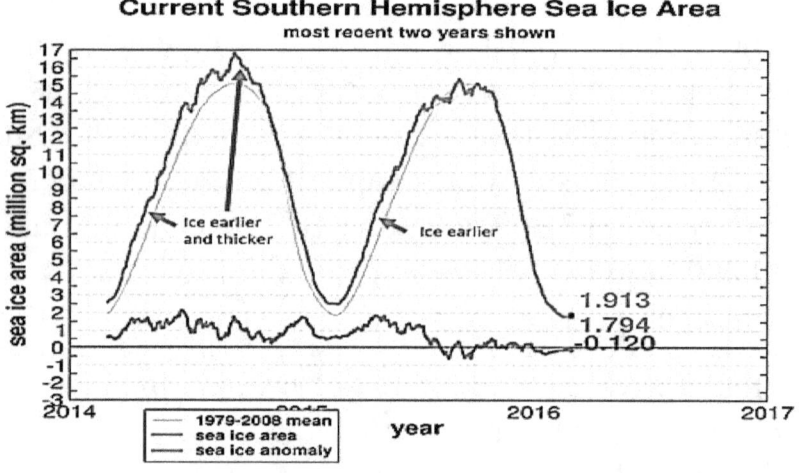

The following graph simply is looking at the coldest days with 2014 and 2015 superimposed.

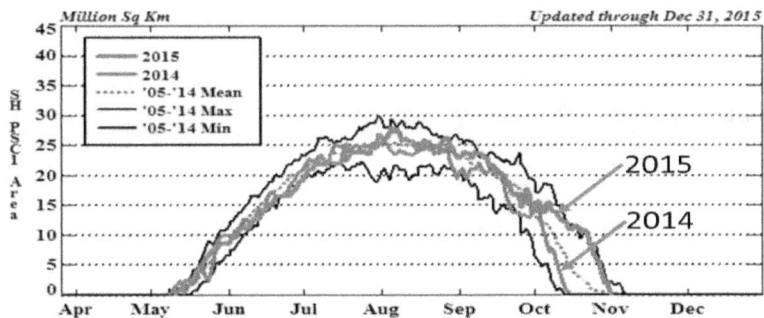

With the coldest October ever recorded in Antarctic in 2015, someone is not being responsible telling the world that there is no hope for civilization and we need to eliminate coal and oil fuels.

After seeing how clouds, might be something to look at, let's do it.

More Clouds and Water

Between 1990 and 2000, a 5% decrease in cloudiness *increased* the total earth surface radiative flux by 6 Watts per square meter. The entire CO_2 forcing as estimated by the UN's IPCC climate panel was just 1.6 Watts per square meter over the past 250 years. While CO_2 can cause warming, its role is tiny compared with natural influences, and much, much lower than the UN's IPCC estimates. Water, water vapor, and clouds are dominant, but here is the rub!

Only CO_2 can be taxed!

Recently the Weather Channel announced that at any one time there were 20 **trillion** gallons of water in the atmosphere over the States at any one time. This equates to 75,000 million tons of water, or 10,000 tons of water in the air for every square kilometer. I am certain the members of our governments taking part in this travesty would greatly desire to tax all that green gas. Just think about we would have to control our waste so that none of the water gets in the air and taking a shower longer than 2 minutes would be severely taxed. The following graph may tell us some important information. We can see that the higher altitude

Stratosphere [the line that starts highest and ends lowest] is much more dynamic and getting cooler while the lower Troposphere is fairly steady and getting slightly warmer showing. Ta-da!!!!! The lower air is different than the Stratosphere line because of-------CLOUDS controlled by cosmic rays.

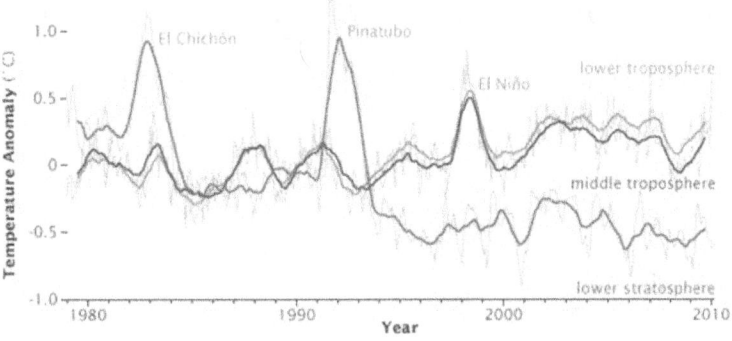

Most Dangerous Greenhouse Gas

The IPCC can't see this simply because they don't consider the main greenhouse gas as water vapor. If *95% of the greenhouse gas is water* and it is; we must attribute clouds as one of the major causes for temperature increase. Let me give you an example. With CO_2 at 390 ppm and water vapor is somewhere between 10,000 and 40,000 ppm and wreaking havoc on our atmosphere. When can we start to be honest and see that if we have been in cooler periods, it is natural to expect that it will get warmer so quickly, there must be a much larger driving force than a trace of CO_2 gas in the atmosphere? The Sun is a major player [man is NOT]. ***Cosmic rays from interstellar space modified by the sun's solar wind*** plus the sun's and earth's magnetic fields are next largest factors after the

113

sun itself. Following this is CLOUDS. Way below all these is what we call non-water greenhouse gases which play a very minor role. Here is something the NOAA didn't tell you.

If the CO_2 is removed from the atmosphere, water vapor will absorb that band of infrared energy and make just as much heat as if the CO_2 had never been there.

As Cloud cover changes so does the Ozone.

More Ozone

Tropospheric ozone [O3] has a short atmospheric lifetime and is a potent greenhouse gas. Chemical reactions create ozone from emissions of nitrogen oxides and water in the presence of sunlight. As the sun gets hotter, we get more Ozone and a cooler sun reduces the Ozone. We like to thing Ozone is great stuff, but in addition to trapping heat, ground-level ozone is a pollutant that can cause respiratory health problems and damage crops and ecosystems.

Ozone Hole

I know you've heard of this stuff and the solar cycle matches the Ozone hole cycles as you would expect, but someone seems to be trying to force fit other things to gain wealth. The image following shows the almost closed hole over Antarctica during the 70s and the larger opening in the 90s [second row]. The last row shows more recent patterns where the opening almost closed again in 2002 and appears to be closing again by 2012. <u>UV light hitting the Earth surface varies by as much as 400%</u> over the solar cycle due to variations in the protective ozone layer. In the stratosphere, ozone is continuously <u>regenerated by the splitting of O_2 molecules by ultraviolet light</u>. During a solar minimum, the decrease in ultraviolet light received from the sun OBVIOUSLY leads to a decrease in the concentration of ozone. This allows increased UVB levels to reach the

Earth's surface. I believe you can begin to see this is a chicken and an egg overview. Let's just say that when the Sun emits more UV, the Ozone hole gets larger and during higher activity, it gets smaller. Get rid of the sun and no more OZONE FEAR.

Polar Ozone Depletion – The "Ozone Hole"

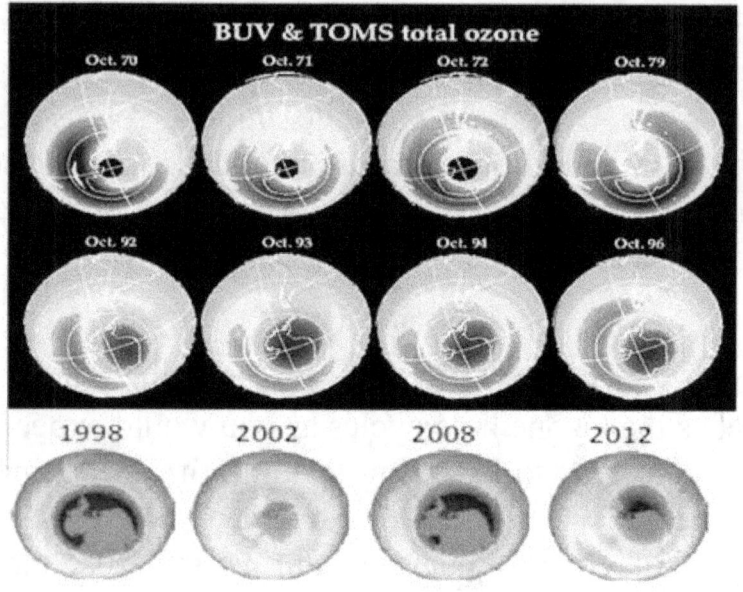

More Temperature Details

Sea Surface Temperatures-Here is another alarming bit of data passed on by those with ulterior motives. The Sea surface temperature as measured by NASA satellite which shows a general downward slope, but that isn't all. The average from 1998 till now has been drifting downwards and since 2009 the temperature has been declining continuously.

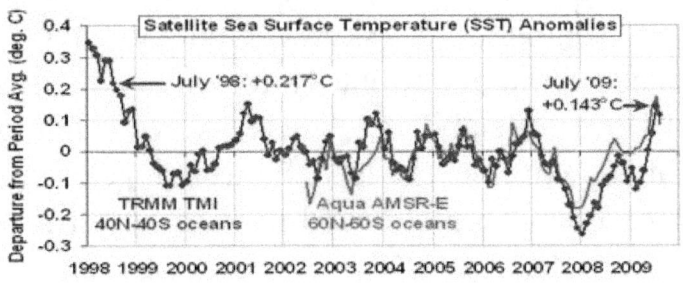

Temperature Trends- If we expand out to the last 4000 years, the Greenland Ice core shows temperatures mostly stayed the same the whole time except for the little dip we started coming out of about 300 years ago. Since that time, the temperature has been trying to get back to NORMAL.

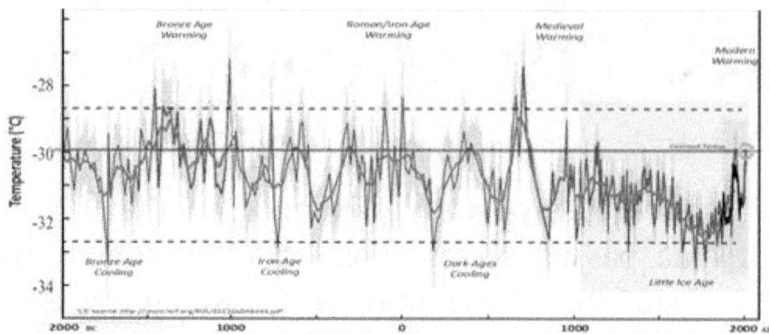

Let's look farther back and see what Greenland's temperature has looked like over the last 10 thousand years as the Earth recovered from a massive shift in its axis. The temperature of Greenland showed the massive shift as temperature skyrocketed to the levels experienced in the current Holocene Age as shown below. Once it stabilized, there were much greater temperature extreme shifts 8 thousand years ago that anything we have experienced in the near term and the Earth did not die.

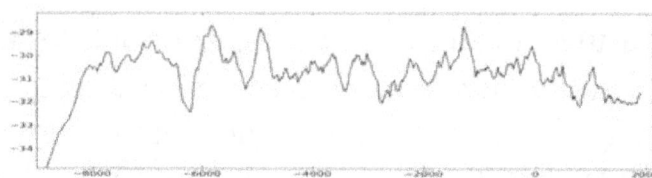

Let me just move back to 11 thousand years ago as the Earth temperature dropped by 8 degrees followed by a 20 degree rise that took 2500 years. [See next chart] Hopefully you can imagine how horrible it was. Mammoths froze in place in Siberia, piles of animals were twisted together in Alaska, and hundreds of feet of fish dried in the mud at Karoo Africa. With all the shifting the atmosphere would have gone crazy. As the atmosphere was in disarray, Cosmic, X-ray, and UV waves would have scorched and challenged cloud formations and the Ozone hole would have had issues, but the Earth survived as did many of the people of that time. From this perspective, it's the calmness of our temperature that seems strange.

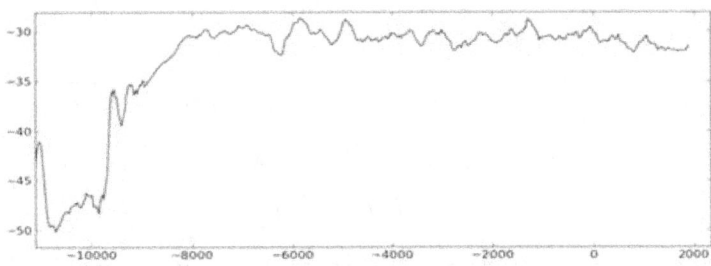

Rather than worrying about building massive solar arrays to generate power for air conditioning, there is another, far more likely, possibility I would worry about. I'm talking about the return to an ice age. As I have tried to show, while extinctions occur every 100 thousand years, interglacial time periods only last for 10 to 20 thousand years and this one started 10 thousand years ago. We could see an Ice Age pretty soon whether we eliminate CO_2 fueled electricity or jet fuel power aircraft, or even the Ionospheric heating by HAARP or EISCAT. We will look at that a little, but we still have to learn more about some of the causes of the Artic overheating while the rest of the planet stays the same or gets colder.

More CO2 Detail

Here we find that CO_2 in the atmosphere went up slowly from 1870 to 1970 (290 parts per million, ppm to 320ppm).

It then, by fantastic fiction, the CO_2 level rose a further 70ppm to about 390ppm in 2010. This would be the vanguard of disaster.

Everyone would melt down. The graph below shows the CO_2 content captured in the Ice Core samples of Antacrctica.

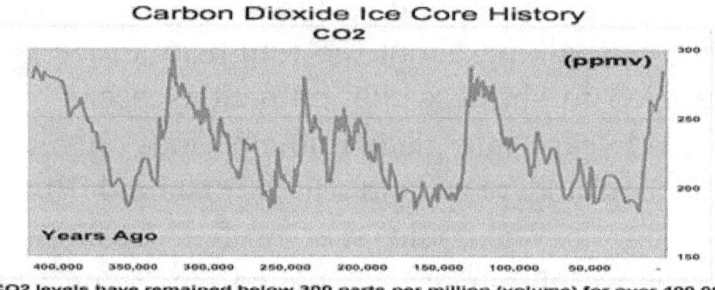

Carbon Dioxide Ice Core History

CO2 levels have remained below 300 parts per million (volume) for over 400,000 years

The following graph shows the faked increase sice 1600 or 1700. While there was a slight increase until 1996, No published Ice Core samples have been made since. All data since them comes from aeresol CO_2 in Hawaii.

I brought this up befrore, but let me show you how it was done. The first graph below shows the transition between CO_2 measurements in Antarctica ending in 1996 and Aeresol CO_2 measurements from Hawaii and from satelite imagry that started about 1960 and is shown with directed data up until 2008.

Notice there is a 33% difference in the absorption levels of CO_2 in Antarctica and the aeresol figures and the slope of increase is 3 times as fast.

Knowing the abosorption level, simply reduce the slope of Aeresol CO_2 by a 1/3 and what you have is shown in the middle graph showing a very slight CO_2 within "NORMAL" levels [I probably should have reduced the slope more taking in account for the huge slope differences of the 2, but this is worst case. Instead of using logic, NOAA used trickery and misdirection to make the last graph that has been scaring everyone, including climatologists wanting there to be global warming to allow them to write papers and gain notoriety.

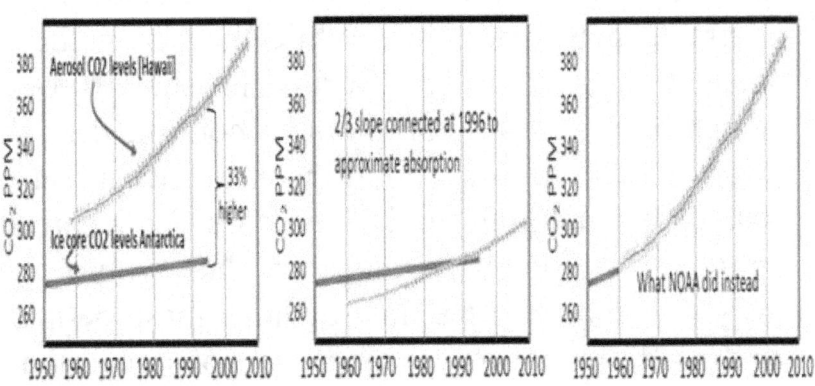

By putting the graphs together to make it look bad, they simply attached the Hawaiian levels onto the

Ice Core data as if 100% of the CO_2 in the Hawaiian air gets to Antarctica and is not recombined before being buried in the Ice.

The problem is we aren't talking about stupid people we are talking about dangerously bad ones. Ignore the name on the graph as it is another lie.

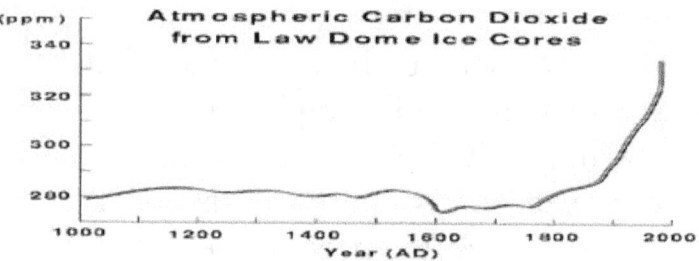

They did the same thing with NO_2 levels and CH_4 levels and they produced terror to reap benefit.

Expanded Data

Have you ever noticed that these "sicientists" never show expanded views of the Ice Core data? There is a simple reason for this. While they need the very ancient data from Ice Coring to show how our temeprature is now going out of control, they don't want you to know about inconsistencies between satellatie imaging and Ice core concenrtration graphing. I thought I would just show you a couple so you can appreciate how devious these people are and so desperate to convince you of their made up disaster. On the left is the data from Greenland Ice core temperature mapping and the satellaite temperature map is shown below it for the same general time. To the left is a similar interaction with data from Antarctica. Please see that there is lost data suppressed by those wanting to show a particular evil wen it doesn't exist at the level they are presenting. The high temperatures from the 1940s is completely missing from the NOAA 2015 data well after the falsehoods I already brought out earlier were uncovered and proven. While the Antarctic data looks more consistent, there still are errors as the downward slope of the Ice core is converted to an upward slope from the NOAA data.

 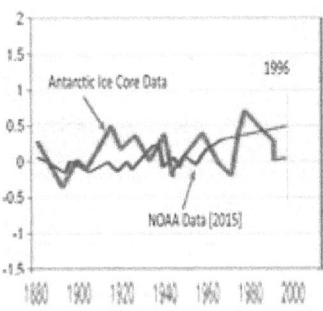

IPCC Data Points To Treachery

Luckily for you, you don't need to take my word for anything. The IPCC told on themselves in a paper presented in 2009 that tried to confirm massive increases in CO2, but what they ended up showing was TREACHERY. Here are some snippets. I have not put them together just to make it look bad, but I didn't think you would want to go through all the mess so I boiled it down a little. The IPCC acknowledged CO_2 has something called a short residence time, stating:

*"The turnover time of CO_2 in the atmosphere is about 4 years. This means that on average it takes only a few years before a CO_2 molecule in the atmosphere **is taken up by plants or dissolved in the ocean**.*

As you read through this, remember the Hawaiian Aerosol CO_2 detection is the thing that is causing the entire ruckus. The data taken in the Hawaiian air included substantial amounts of CO_2 that would be absorbed into plants and never get into the Antarctic Ice.

"The CO_2 response function used in this report is based on the revised version of the Bern carbon-cycle model used in Chapter 10 of this report. --About 50% of a CO_2 increase will be removed from the atmosphere within 30 years and a further 30% will be removed within a few centuries. The remaining 20% may stay in the atmosphere for many thousands of years".

Let me first say that the 30% that won't mix with Ice for centuries should have been used in the amalgamated data. Let me just say if 20% CAN'T ever get into the Ice Core HOW IN THE WORLD could some crackpot just add it in? I could tell you why, but I would have to spit a lot. Let's continue.

"A quasi-equilibrium amount of CO_2 is expected to be retained in the atmosphere by the end of the millennium that is surprisingly large: typically ~40% of the peak concentration enhancement over preindustrial values (~280ppmv).

Here they are trying to say, ignore what I just told you ---never mind, I have no idea what they are saying it is complete gibberish, but we have to continue.

"If the partial pressure of CO_2 varies and the hydrogen ion concentration was kept constant, the relative changes would be the same in the sea as in the atmosphere. As the total amount of CO_2 in the sea is about 50 times that in the air then practically all excess

CO$_2$ delivered to the atmosphere would be taken up by the sea when equilibrium has been established".

Oops, I don't think they were trying to say, the Aerosol numbers had no effect on the Ice Core samples. Maybe I don't understand their English.

Estimates of past carbon dioxide concentrations derived from ice cores drilled at Vostok, Antarctica and Siple Station, Greenland are combined with the modern instrumental record from Mauna Loa Observatory to show the relationship between atmospheric CO$_2$ changes associated with ice ages and the modern increase in CO$_2$ associated with human activities. Natural control of atmospheric CO$_2$ ended at the time of the Industrial revolution, when humans began burning fossil carbon fuels, <u>manufacturing cement</u>, and removing forests at an increasing rate.

This makes you wonder if they read their report before telling the world that they were going to ignore it and everyone else should as well and they should somehow understand that all the CO2 in Hawaii gets to the Ice cores somehow. By the way, this report is the first time I have heard of an environmental group being mad that we have cement houses.

Please do not tear down your cement home and build it out of straw to protect the Earth.

Another CO_2 Anomaly

Something else doesn't add up. If the massive increase in CO_2 charts were in anyway correct and CO_2 was supposed to be the harbinger of global disaster. While CO_2 was increasing slowly over that total period, global mean temperatures actually went <u>down</u> *twice* in that period; first between 1870 and 1915 and then again from 1940 to 1970. To make things even weirder, temperatures actually went up rapidly in the 1915 to 1940 period when the change in CO_2 was almost flat. Luckily for those with stock in windmills, there was **one** period (1970-2010) that showed similarity as CO_2 and temperature both increased. With 3 out of 4 periods **not** matching, it is inappropriate to think there is a correspondence between CO_2 and temperature? These trends are shown in the following figure. We can only conclude that there is a poor correlation between CO_2 and earth's temperature, but Al Gore still got a Peace Prize for ignoring most of the data. [By the way, Mr. Gore actually shared his Peace Prize with the IPCC for their part in this scam.] Later we will look at something that does track, but it is not controlled by forcing people to drive electric cars. The following chart shows wide variations in Arctic temperatures [the top line] with almost no change in CO_2 [bottom line] until a couple hundred years ago when CO_2 levels began rising and the Temperature stayed almost stationary. No one in their right minds can tie CO_2 to temperature fluctuations. This is just one example.

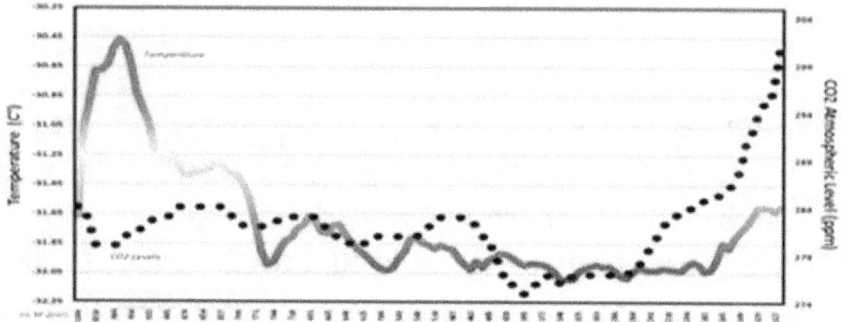

Today we have an even larger differential as shown next. The average temperatures have settled and are slightly decreasing while the CO_2 levels are skyrocketing according to Greenland data 1997 thru 2012 as shown below.

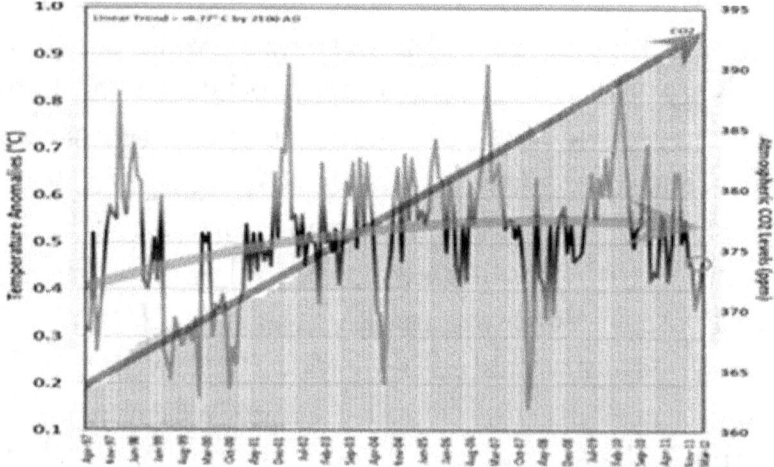

The empirical evidence is so overwhelming that even the vast majority of conspiracists climate scientists (over 97%) agree that the predicted "accelerating"

global warming has been non-existent over the last 15 years.

Let's hypothetically say CO_2 is changing our temperature. The charts below show both CO_2 levels and temperature captured in the Ice Cores from Greenland and Antarctica. The one on the left is from Antarctica over the last 50 thousand years. The thin line that begins below the erratic temperature curve shows something interesting. CO_2 doesn't change until after temperature changes as temperature controls CO_2 rather than the other way around. The second graph is from Greenland in case we didn't see things right, we see that Temperature, the erratic line changes well before CO_2.

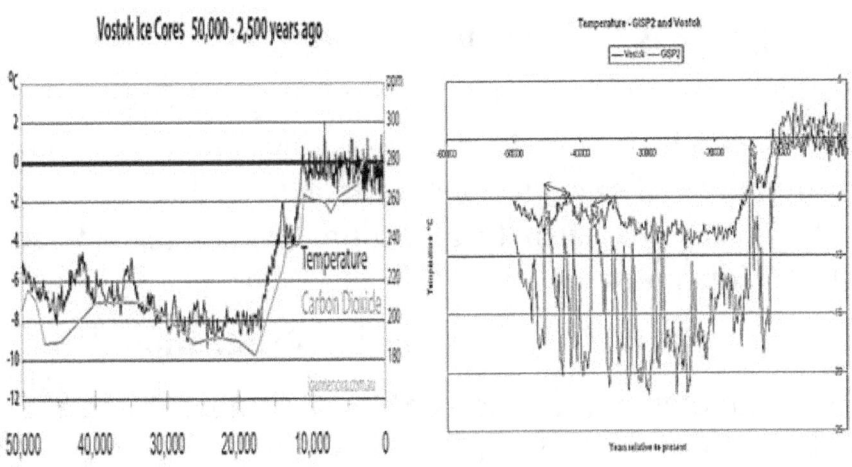

Attack Everything

To twist the truth, to build horror for profit, to ignore the obvious, and to bully the people of this planet so limit concern over some testing that is kept secret simply because they do not exactly know what may happen might be considered noble by some, but my ideas are somewhat different. Here are snippets of false concern.

Ignore Real Greenhouse Gas Water Vapor

We could go on and on as new "Greenhouse Industries" to eliminate water production would make millionaires of the proponents that saw this horror as water heats our earth to alarming levels. Instead let's attack Volcanoes.

Volcano CO_2 and Sulfate Production

For completeness, let's look at these things. If you were looking for a CO_2 emitter you would be on the right track. Volcanoes are active around the world and continue to emit carbon dioxide as they did in the past at the rate of 130 to 230 million tons of carbon dioxide per year but burning what are called fossil fuels, releases about 26 billion tons of carbon dioxide, into the atmosphere every year so we can disregard volcanoes as well.

Sulphates have been measured in samples from the Greenland ice cores. A general description of what was found is shown below from 1880 until about 2000.

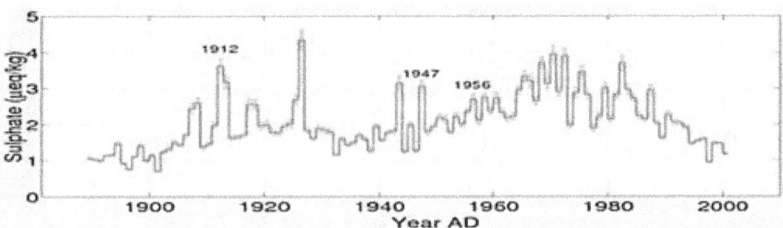

Notice that volcanoes push out Sulphates. The volcanic imprint is seen in the early part of the record: Katmai (1912), Hekla (1947) and Bezymianny (1956) all caused elevated levels of sulphate, but during the 1940s through the 1970s, the chart shows that something was still lingering in the air and finally began to be deposited at elevated levels. We can believe most was from the remains of the predecessor volcanic action. In this period, sulphate pollution caused acid rain and severe damage to trees in some areas. Some believed that sulphates should all be falling to the ground overnight and the bubble was from massive fossil fuel automobiles of the 70s. The bubble dissipated between 1970 and 2000. Some claim the reduction was the elimination of Sulfates in automobile emissions. I believe this is a good thing as I like less acid rain, but the timing suggests there were other dissipaters like fewer eruptions at this time.

Nitrous Oxide

Nitrous oxide is produced through natural and human activities, mainly through agricultural activities and natural biological processes. Fuel burning and some other processes also create N_2O. Like the other claims, concentrations of N_2O have risen approximately 20% since the start of the Industrial Revolution, with a relatively rapid increase toward the end of the 20th century. Here is what you were told. Overall, N_2O concentrations have increased more rapidly during the past century than at any time in the past 22,000 years. The big problem with that statement is that we do not know what the airborne Nitrates were back then. All we have are the levels that got into the ground and were frozen. The slight increase from 1880 until 2000 is shown below left.

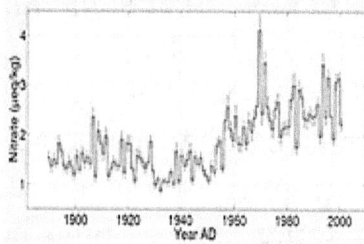 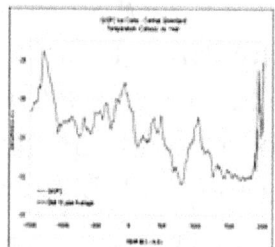

Lead Pollution

The Chart above right is one of those where they used the Ice Core data for everything before 1990 and then switched to Aerosol data making it look like wild fluctuations over recent times. Please don't look at the last part. All these values are very tiny and have been blown up to make them look significant, but there are only tiny amounts in the air.

Black Carbon

Black carbon is sort of a new worry. It is a solid particle or aerosol, not a gas, but some tie it also to warming of the atmosphere. It is true that black carbon particles can directly absorb incoming and reflected sunlight in addition to absorbing infrared radiation very slightly, but certainly not as good as gases that can actually bond with other air particles. Here is the worry. This black carbon can be deposited on snow and ice, darkening the surface and thereby increasing the snow's absorption of sunlight and accelerating melt. If you do see Ice getting black, please scrape it a little so the Arctic won't melt. This will not slow down the periodic melt cycles and it is nothing to worry about. Don't get me wrong. Having too many carbon particles in the air can make smog and smog can kill. We certainly should reduce carbon particles in the air in urban environments----but that has nothing to do with the Earth overheating!!!!!!!!!

Hot, Hot, Hot

Our Atmosphere took and abrupt change alright, but not because of CO_2. Models no longer work and haven't worked since 2003 according to climate scientist Kenneth Trenberth and others.

Kenneth Trenberth, *climate scientist: "Until 2003, scientists had a reasonable understanding where the sun's trapped heat was going; it was reflected in rising sea levels and temperatures. Since then, however, heat in the upper ocean has barely increased and the rate of sea level rise slowed,...they put forward a climate model showing that decade-long pauses in temperature rise, and its attendant missing energy, could arise by the heat sinking into the deep, frigid ocean waters, more than 2,000 feet down."*

The reason models don't work and CO2 is climbing while temperatures drop, and the Oceans are not flooding the lands and we are seeing record low temperatures and the return of Ice to the Artic is not because CO2 comes and goes willy-nilly.

Contrails to Chem-trails

About this time the condensation trails [Contrails] behind passenger jets started looking strange. Contrails are simply vapor trails which are left in the wake of a sky borne jet airliner. They appear at the very rear being ejected as the exhaust from the jet airliners, and dissipate quite quickly depending on the ambient atmospheric temperature, pressure and humidity at the time. Contrails are random and generally depicting normal traffic flow of passenger liners. If a crisscross pattern is ever seen, it is not from commercial airliners.

Straight parallel lines and fumes coming from the tip of the wings are not Contrails.

Contrails cannot be generated from the tail of the airliners as shown below.

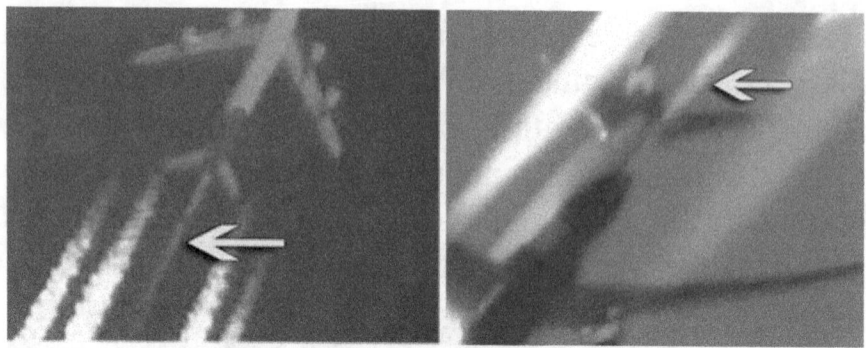

Contrails cannot stop and start again as they are expelled unless someone turns of the jets. That would be dangerous.

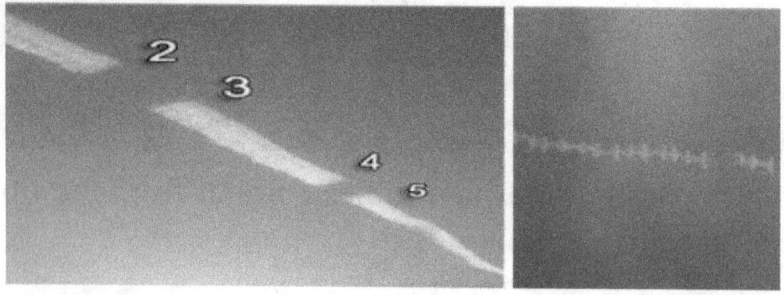

Contrails are not laid down by commercial jets flying in parallel as shown below. Very seldom do they have commercial jets taking off at the same time for the same place.

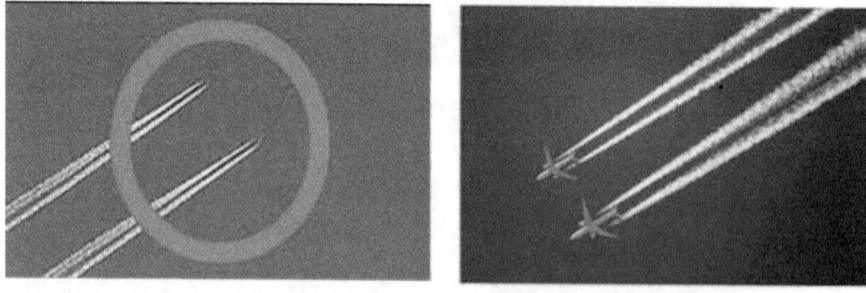

If some of the streams are dark and some are white, there must be a difference in how the two streams are generated. Some might believe one has metal oxides mixed in as shown below.

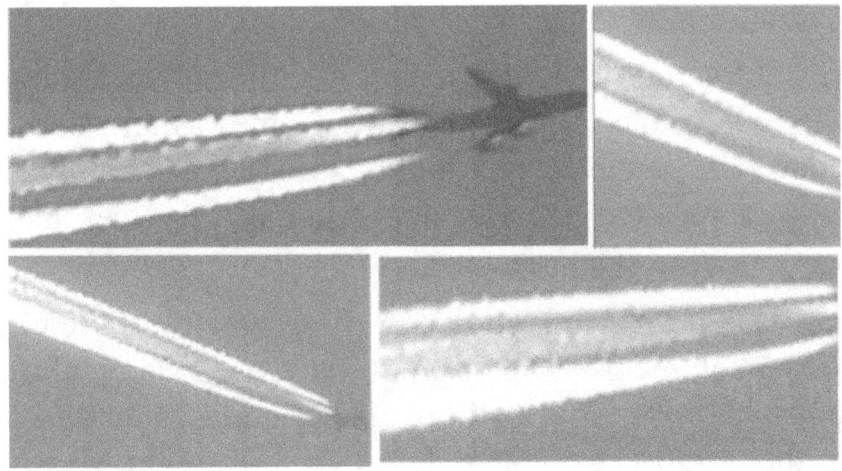

You can really tell when someone forgot to hide one of the modified jets used for dispersion, the emission ports were found all over the thing.

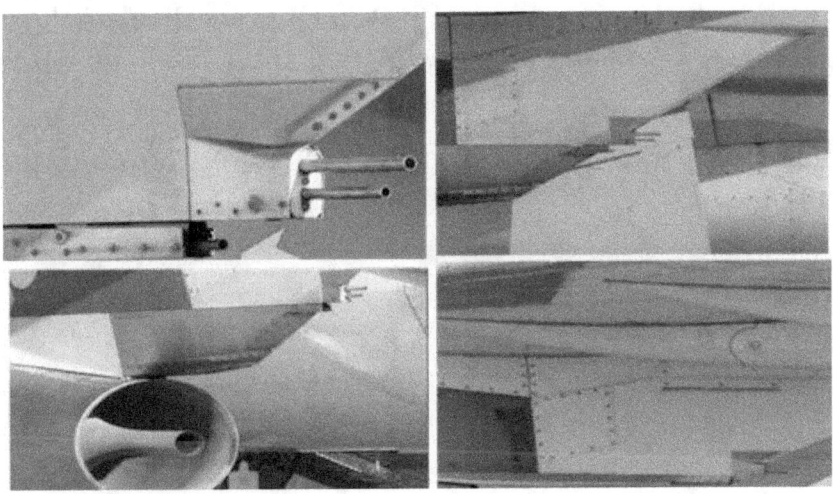

In 1995- The US Air Force published a paper about their research on weather modification using cloud seeding.

In 2008-Cloud-seeding using Chem-trail technology was used at the Beijing Olympic Games as an effort to reduce the terrible air pollution problem, certainly no outrage resulted from trying to make us feel more comfortable, so let's get back to the Ionosphere Heaters.

Increase in Storm Activity- Before we look at the next charts, let me re-establish a timeframe. Many of the Ionosphere Heating Stations of the world came on line in the 1990s. Certainly HAARP did. The Massive antenna array that allowed it to reach 1.5 Giga watts of energy wasn't completed until 2006, but massive amounts of testing was carried on from According to the chart following, since 1993 until now North Atlantic storms have increased by 50% because of ~~global warming~~ something. Did anyone at NOAA consider there was a correlation much more reasonable than CO_2?

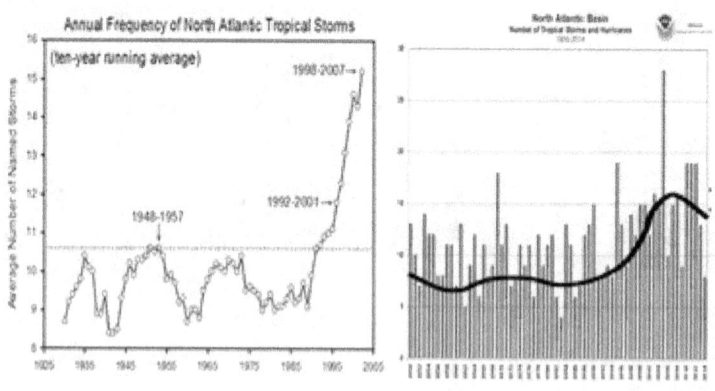

The first chart shows a running average of North Atlantic storms from 1925 until 2007 while the second chart shows a similar detail from 1950 until 2014. We see a rapid rise from 2000 until 2007 followed by a reduction until today. What we also find is that South Atlantic storms are not increasing like the northern hemisphere as if someone is messing with weather in the North and still the NOAA team try to force a tax on burning coal.

HARP and Greenland Ice-Before we look at the next images, let's understand that 1990 was before HAARP, 2007 was the year the IRI came on line and the reported temporary shutdowns of 2013/2014. The images below show the Ice cap in 1990, 2007, 2013 and 2015. From the reduction in 2007/2008, the Ice Caps have been almost steadily increasing to over 63% larger. We can believe 2007/2008 were a banner years for HAARP research, but a devastation for the Arctic Circle. While I don't have to mention it, NOAA saw nothing similar.

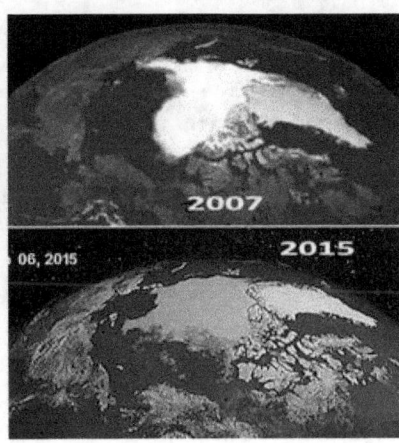

The image next shows low ice levels in the Arctic and increases of Ice in Antarctica at the same time in September 2012.

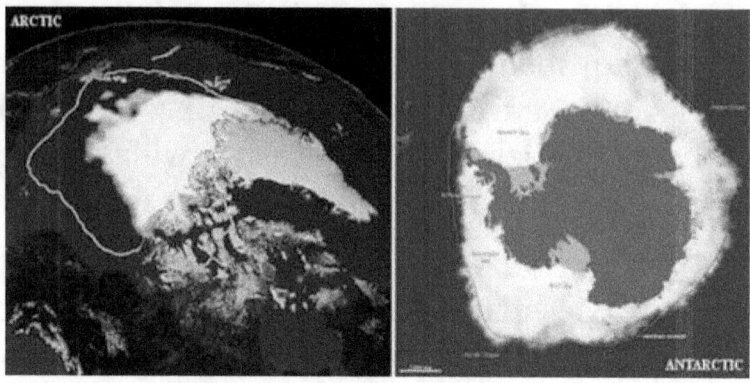

If we compare the Ice levels in May 25 1995 and that noted in May 25, 2015 we see a massive resurgence of ice as less HAARP experiments were acknowledge during an attempt at closing it down.

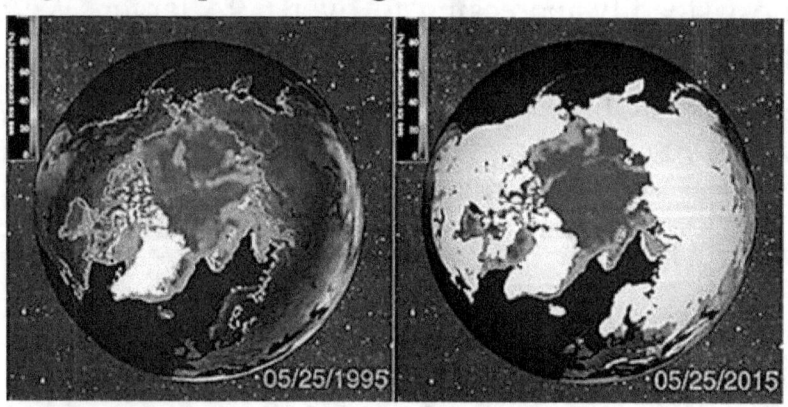

CO$_2$ Cannot Absorb Sunlight

We are taught in school that about 50% of the sunlight is absorbed in the air and described below in the atmospheric absorption bands of the various gasses. [30% Oxygen, 20% water, no CO$_2$, and no N$_2$O].The earth reradiates the energy at a higher wavelength and about 30% is absorbed again. [14% water, 10% Oxygen and 4% CO$_2$ and 2% N$_2$O] but this radiation is about 10% of the sun's original emission through the atmosphere. Of the total energy, Oxygen absorbs 28%, water absorbs 20%, methane, CO$_2$ and N$_2$O all together cannot absorb more than 0.6%. Therefore, our best bet is to reduce either water or oxygen as none of the greenhouse gasses that don't make things green can do anything. In the aggregate we find absorption of solar and reradiated energy is 57% from oxygen, 41.1% from water, and 1.8% from every other gas together.

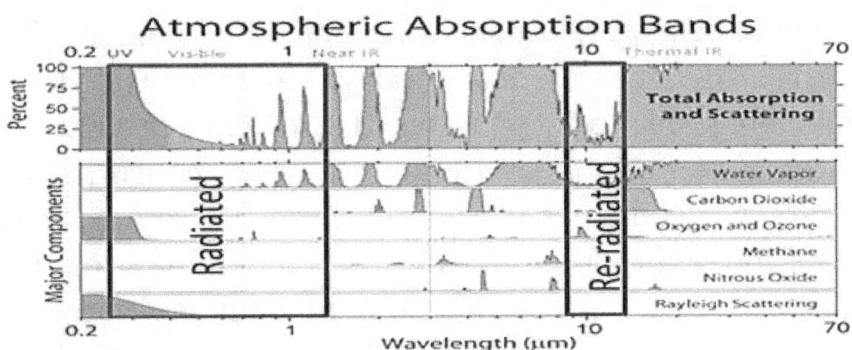

The following absorption band chart shows both the solar radiation and the terrestrial re-radiation levels and the materials that can absorb the energy. CO_2 is not a good absorber.

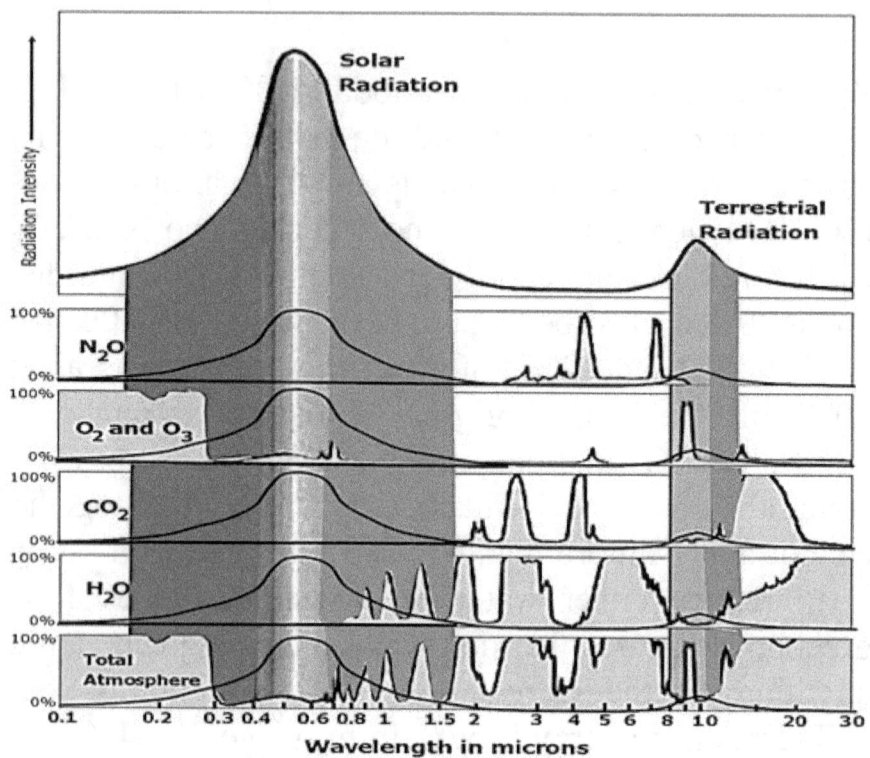

The Last Ice Age-From 1200 until 1700 the average temperatures were about 4 degrees Fahrenheit [2.2 degrees C.] colder than today. All of a sudden, something sparked a massive rise in temperature and over a matter of about 40 years; we were about what we are today. Don't get me wrong; I'm not saying there were too many jets in the sky, automobiles on the highways, or too much electricity in the homes, because I'm not an idiot. Today, CO_2 destruction conspiracists look at a tiny rise of ¼ of a degree and publish a new book of the Destruction of the Earth. They develop another failed environmental change model and they put a higher bounty on the elimination of the economic security of the coal producing States. The average hemisphere temperature over the last 10 thousand years is shown below.

Northern Hemispheric temperature reconstruction for the past 10,000+ years

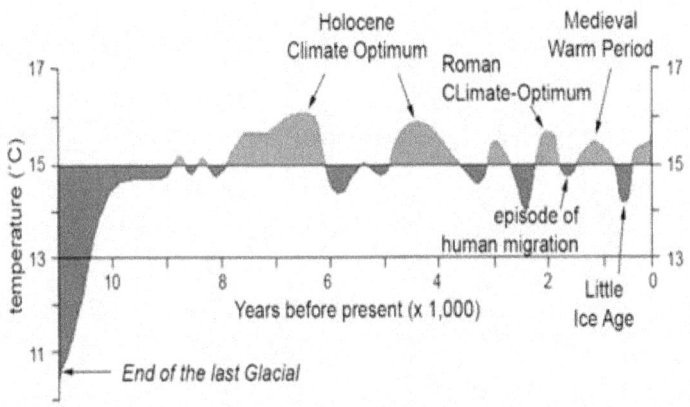

Clouds Track Temperature -I want you to look at two graphs together for a minute. One is the temperature and the other is the water vapor cloud cover average for the world starting in 1983. As the Cloud cover continuously reduces until about 2000 the temperature rise and when the cloud cover percentage remains constant, the temperature stays about the same. Because everyone knows that water vapor can easily absorb solar energy, this makes sense so the UN's Intergovernmental Panel on Climate Change [IPCC] pushed CO_2 as a culprit with absolutely no reasoning until we read ominous news.

143

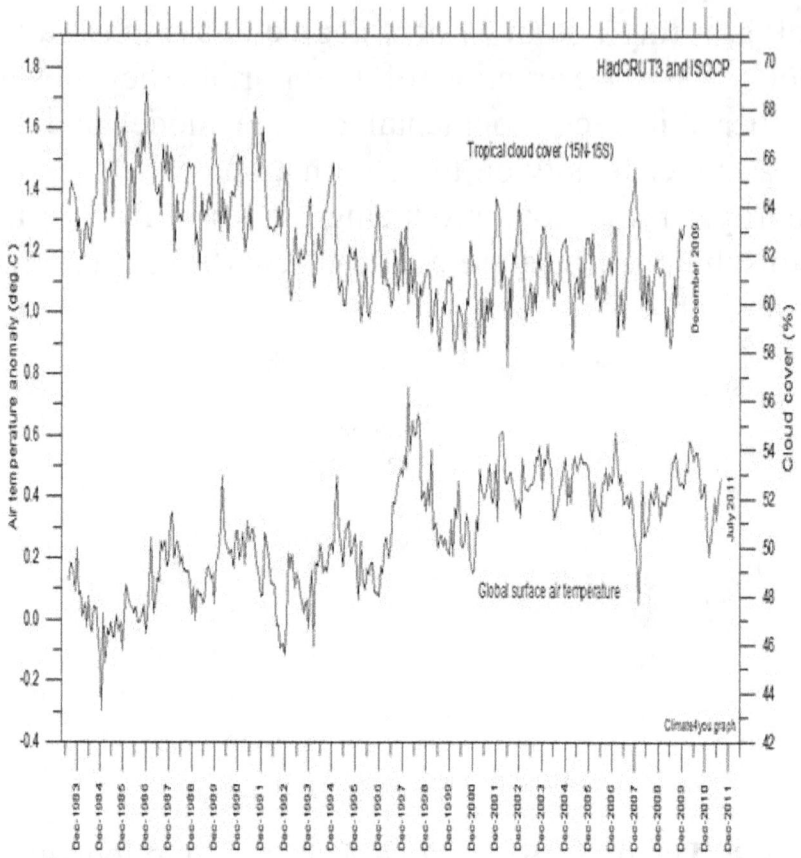

I know it doesn't sound like the Arctic could be heated by most of the things that are being blamed for the heating and I think if we just look at the air we will see something interesting.

Air Tells Us the Truth

The image below left shows the normal airflow patterns around the world identified as Polar and subtropical Jets.

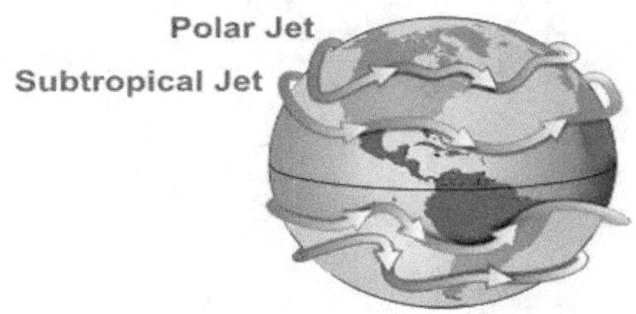

From the Climate Change Institute we can look at the wind pattern from May 14 [middle] and November 14 2014 [right] and see something peculiar around both the HAARP site and the EISCAT equivalent.

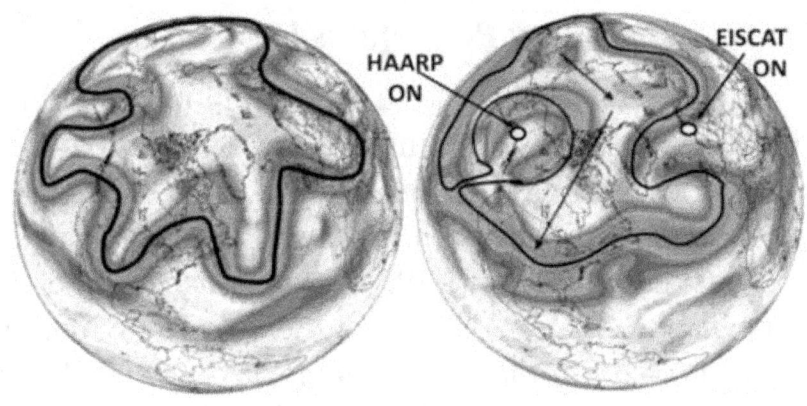

It appears that during May they didn't transmit and in November they did. We begin to see a pattern. The January 2014 weather map from NOAA shows a noticeable anomaly around the EISCAT Ionosphere Heater while in January 2015 below it we see it is not transmitting. We can believe many know what issues are arising from the experiments.

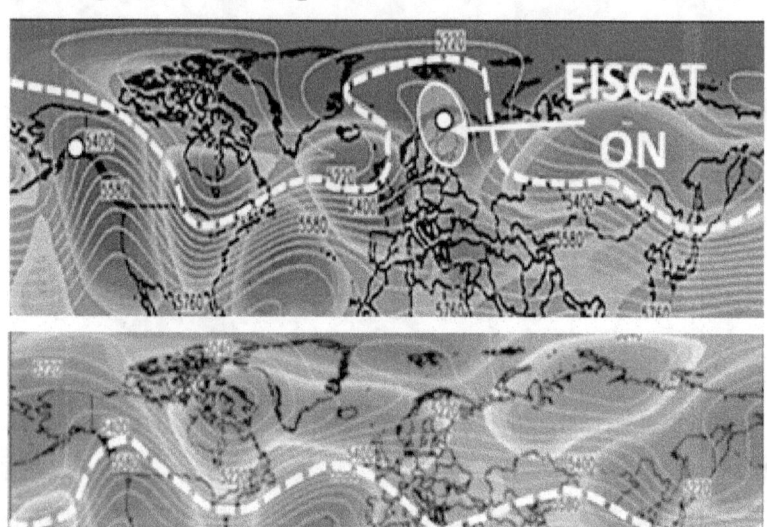

Just How Powerful is EISCAT

As I mentioned HAARP has a sister in Norway and that is the one that seems to be moving high pressure around on the preceding images. EISCAT (European Incoherent Scatter facility in Norway has been heating the Ionosphere since 1979. On February 1996 the European Space Agency calmed viewers by explaining that *the DASI Digital All-sky imager of the EISCAT showed and auroral arc located slightly south of the facility during 2 ionosphere heating cycles. A development of spiral like forms occurred after the EISCAT was turned on.* Not only could this thing make its own Auroral Borealis like the HAARP system in Alaska, we can certainly believe it disrupted weather patterns. It power is on display in the next images.

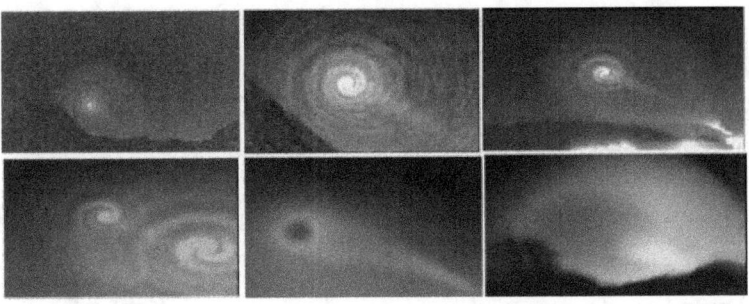

The next image shows how plasmas can be built from the intense transmissions. Notice the cavitation rings of this plasma generated by EISCAT.

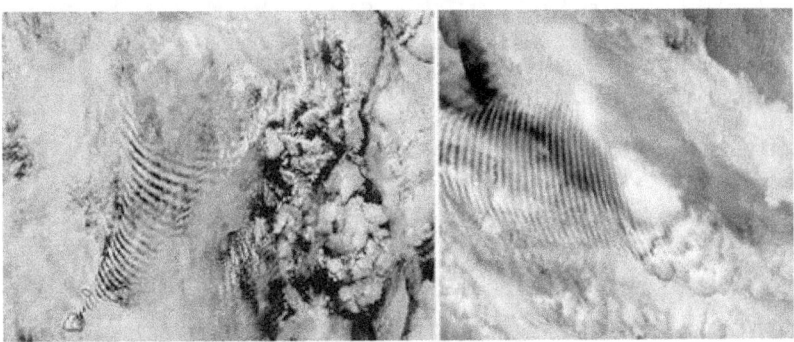

HAARP Concerns

I suppose you can tell either HAARP or EISCAT are at the center of most weather modification conspiracies. While there are many experiments going on at both locations. They are thought to be a secret weapon in weather modification schemes. The Middle East is especially going after the EISCAT group after having record high temperatures and drought the last couple of years. Since one of the main purposes for both is Weather research, it is not difficult to believe that many times they simply go too far in their blasting of the various layers of the Ionosphere and into the Magnetosphere. I think most should agree that some of the research is reasonable and some of it is less than helpful to our environment as major group began questioning how HAARP and EISCAT could be affecting Global Warming.

EU Concerns-In January 1999, the European Union *called the various Ionosphere heating projects a global concern and passed a resolution calling for more information on health and environmental risks.*

Secretary of Defense Concerns-At a news briefing in 1997, the Secretary of Defense, William S. Cohen, made a statement to the media where he outright admitted that there *are* environmentally-altering technological threats in existence. Cohen's comment at that briefing has provided fuel to the conspiracy theorists' claims that technologies like HAARP *could* be manipulating the weather or exploiting communications. Cohen's words are published in news briefing transcripts on the U.S. Department of Defense website, and include the statement: *"Others are engaging even in an eco-type of terrorism whereby they can alter the climate, set off earthquakes, volcanoes remotely through the use of electromagnetic waves."*

EU Committee on Foreign Affairs-Two years later, a report from the EU Committee on Foreign Affairs, Security and Defense Policy stated: *"Despite the existing conventions, <u>military research is going on environmental manipulation as a weapon</u>, as demonstrated for example by the Alaska-based HAARP system."*

Besides the real weather concerns, the HAARP has been blamed for Hurricanes Katrina, Sandy, Wilma, and Rita. It has even been said that the organization is actively practicing mind control on American citizens with shortwave radio signals. Please don't be one of these people. It simply is not good for you. There is no question that if you mess with weather in one part of the Earth, other parts will be affected, but it would be difficult to believe these efforts are on purpose except for possibilities from the miniature Puerto Rican Ionosphere Heater.

2014 Storms-If you remember, I told you that Puerto Rico had one of the Ionosphere Heating stations. One would think that is a long way from the Auroral Oval, but what if one could direct storms a little. It would not take much to build a small high pressure area that could nudge storms. Possibly even a small Ionosphere heater system could redirect storms. I haven't looked into this much, but who knows. The image below shows the major storm paths of 2014. I'm not getting into this much more except to say if anyone sees a crisscross pattern of chem-trails along the coastline of Florida they might be on to something. I will say this. If the Puerto

Fico device can push these storms slightly, just imagine what the HAARP and EISCAT can do, especially if they get a little help from those metallic discharges from jets we call chem-trails.

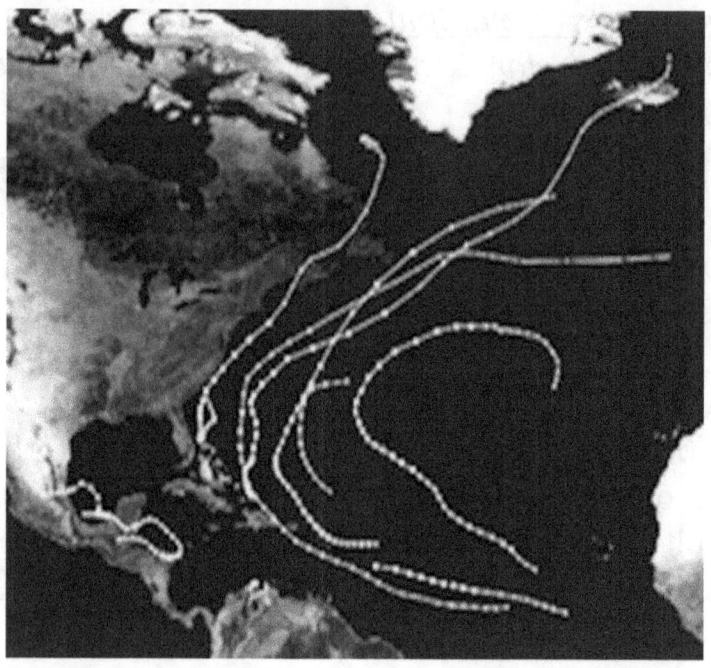

ELF Dangers

If you remember form the beginning, one of the main things being done with the HAARP and similar ionospheric Heaters is to modulate signals using Extra Low frequencies between 0.1 to 20 Hz. These are called infrasonic or infratonic, depending on how information is carried, but, essentially, I wanted to let you know the types of things they are finding out about these things.

We Don't Understand Them Even Today

Disasters are preceded by ELF waves- Animals have been known to perceive the infrasonic waves going through the earth by natural disasters and can use these as an early warning. During the 2004 Indian Ocean earthquake and tsunami, animals were reported to flee the area hours before the actual tsunami hit the shores of Asia. It is not known for sure if this is the exact cause, as some have suggested that it may have been the influence of electromagnetic waves, and not of infrasonic waves, that prompted these animals to flee.

Natural ELF signal of the Earth-Research in 2013 by Jon Hagstrum of the US Geological Survey suggests that homing pigeons use low frequency infrasound to navigate.

ELF has a strange effect of People-One study has suggested that infrasound may cause feelings of awe or fear in humans. It also was suggested that since it is not consciously perceived, it may make people feel vaguely that odd or supernatural events are taking place.

Australia an Scientists Worry-A scientist at the Sydney University Auditory Neuroscience Laboratory stated that there is a growing evidence that infrasound may affect a few people's nervous system by stimulating the vestibular system and this has shown in animal models an effect similar to sea sickness.

Iranian Scientists Worry-In a study of 45 people, Tehran University researchers stated: "Despite all the good benefits of wind turbines ... this technology has health risks for all those exposed to its sound" — in particular, sleep disorder.

Japanese Scientists Worry-In another study by researchers at Ibaraki University in Japan said the EEG tests showed the brain function showed that the infrasound produced by wind turbine were "considered to be an annoyance to the technicians who work in close to a modern large-scale wind turbine

UK Scientists Worry-On 31 May 2003, a group of UK researchers held a mass experiment where they exposed some 700 people to music laced with soft 17 Hz sine waves played at a level described as "near the edge of hearing", produced by an extra-long-stroke subwoofer mounted two-thirds of the way from the end of a seven-meter-long plastic sewer pipe. The experimental concert [entitled *Infrasonic]* took place in the Purcell Room over the course of two performances, each consisting of four musical pieces. Two of the pieces in each concert had 17 Hz tones played underneath. In the second concert, the pieces that were to carry a 17 Hz undertone were swapped so that test results would not focus on any specific musical piece. The participants were not told which pieces included the low-level 17 Hz near-infrasonic tone. The presence of the tone resulted in a significant number (22%) of respondents reporting anxiety, uneasiness, extreme sorrow, nervous feelings of revulsion or fear, chills down the spine, and feelings of pressure on the chest.[1]

Brown Note Worry-The name is a metonym for the common color of human feces. Frequencies supposedly involved are between 5 and 9 Hz, The **brown note** is a hypothetical infrasonic frequency that would cause humans to lose control of their bowels due to resonance. While it was reportedly accomplished once, attempts to demonstrate the existence of a "brown note" using sound waves transmitted through air have failed.

Space Program Worries-The U.S. space program, worried about the harmful effects of rocket flight on astronauts, ordered vibration tests that used cockpit seats mounted on vibration tables to transfer "brown note" and other frequencies directly to the human subjects.

Infrasonic Weapons-Very high power levels of 160 dB were achieved at frequencies of 2–3 Hz. Test frequencies ranged from 0.5 Hz to 40 Hz. Test subjects suffered motor ataxia, nausea, fear, anxiety, or depression, as well as biological symptoms like nausea, vomiting, organ damage, burns, or death—depending on the frequency and power level. Most of these weapons function between the frequency range of about 1 Hz to 30 Hz. also known as Extremely Low Frequency (ELF).

French Weapons-In 1972 France was using infrasonic generators which operated at 7 Hz on its civilian population to break up crowds.

Russian Weapons-In the early 1990s Russia had developed a 10 Hz ELF modulator capable of targeting individuals over hundreds of meters, causing pain, nausea, vomiting, and sometimes death.

American Weapons-US DoD Infrasonic weapon development interest has been substantial since in ELF frequency ranges [7 to 20Hz]. Such a device was supposed to target the brain; changing brain chemistry to influence thoughts and emotions. It also has shown to cause fatigue, pressure in the ears, visual blurring, drowsiness, imbalance, disorientation, vibration of internal organs, severe intestinal pain, nausea, vomiting, liquefying of the bowels, resonating internal organs, causing pressure in the chest, choking, causing irregular breathing and respiratory incapacitation, and even causing death.

Earthquake Fears-According to the *"Acoustic Weapons Prospective Assessment"* article, which

154

appeared in the volume 9, 2001 issue of *"Science and Global Security"*, infrasound can produce <u>localized earthquakes</u>.

The Vortex Launcher, (also called the Vortex Canon, Wind Canon, and Shockwave Weapon), is capable of transmitting an invisible whirlwind of force to effect a considerable blunt impact on a target. It can also be used to transmit chemical irritants to a specific individual or group. It will allegedly be used to <u>disable or destroy personnel</u> such as enemy combatants or disruptive crowds. Most of the current information pertaining to this technology is classified. But in the late 1990s the US Military is said to have developed a vortex gun in conjunction with various defense contractors, the US Army Research Laboratory, and Pennsylvania State University.

Nazi Weapons-World War II a Nazi scientist, Dr. Zippermeyer, invented a device known as the Wind Cannon intended to shoot down Allied bombers with ELF.

Navy Weapon-The Navy is working on a subsonic vestibular [inner ear] weapon to cause disorientation, confusion, extreme motion sickness, and vomiting. through walls

Windfarm Disasters-Windfarm subsonic emission studies showed a consistent and progressive 70% loss in ability to develop concentration , had mental excitement [wandering] followed by mental dullness and finally mental sinking. Additionally was noted head and chest pressure and pain, and even intense pain, heart palpitations, nausea, stomach pain, dry retching, breast pain, dizziness, loss of self- confidence, depression, mania, anger, and heightened emotionality and crying. Long Term exposure studies show <u>increases in heart disease, stroke, cancer, epilepsy, rage reactions, and suicide</u> according to the Alves-Pereira study [1999]. We now hear that dozens of whales beach themselves as they come near these massive infrasonic resonating things.

What Happens in the Brain?- Studies have shown ELF levels of Infrasound causes a variety of psychological effects depending on the frequency and power level. It can cause loss of concentration, disgust, apathy, sadness, depression, fear, anxiety, and panic attacks. The following shows some of the findings. While there are specific frequencies that cause each of these affects, I'm not going to go into that detail in this book. My main objective is to show that the frequencies that HAARP says it tests with are considered dangerous to people and we don't know what happens to our delicate environment. Anyway many are studying the brains so here are some findings.

Epsilon Waves [those less than 0.5Hz.] have caused the following: extraordinary states of consciousness, high states of meditation, ecstatic states of consciousness, high-level inspiration states, spiritual insight, out-of-body experiences, yogic states of suspended animation.

Delta Waves [those 0.2 to 4 Hz] have caused the following: Adults are driven to slow wave sleep, it forces irritability , confusion, or disorientation , is linked to boosting intuition ,linked to cultivating "psychic skills", allows us into "universal knowledge" , provides access to *external* unconscious material, causes deep sleep, lucid dreaming, increased immune functions, hypnosis, decreased awareness of the physical world, deep dreamless sleep, trance, suspended animation, anti-aging. reduces cortisol and increases DHEA and melatonin, provides intuition, empathetic attunement and instinctual insight, miracle type healing, divine knowledge, inner being & personal growth, rebirth, trauma recovery, "one with the universe" experiences, near death experience, and blissful "being" states.

Theta Waves [those 4 to 7 Hz] have caused the following: Young children experience drowsiness, adults experience arousal, deep relaxation, meditation, increased memory, focus, creativity, lucid dreaming, hypnagogic state, recall, fantasy, imagery, creativity, planning, dreaming, switching thoughts, zen meditation, drowsiness; access to subconscious images, deep meditation, reduced blood pressure, cuts down on mental fatigue, increases sex drive, intuitive augmentation, profound inner peace, transforming unconsciously held limiting beliefs, physical and emotional healing, purpose of life exploration, inner wisdom, faith, some psychic abilities, retrieving unconscious material, bursts of inspiration, twilight sleep learning, reverie, high levels of awareness, vivid mental imagery, <u>hypnopompic</u> and <u>hypnagogic</u> states, military remote viewing, activates the repository for memories, emotions, and sensations.

Alpha Waves [those 8 to 12 Hz] have caused the following: cause or are caused by relaxation, cause or are caused by meditation, "super learning", positive thinking, conducive to creative problem solving, stress reduction, intuitive insights, inspiration, motivation, daydreams, lucid mental states, pleasant drifting feelings, mental resourcefulness, tranquil state of consciousness, primarily with pleasant inward awareness, and can cause epileptic seizures.

Beta Waves [those 12 to 30 Hz] have caused the following: alertness, anxious thinking, analytical problem solving, judgment, decision making, processing information about the world around us, increased mental ability, focus, good for absorbing information passively, treatment for hyperactivity and mild autism motivation, arousal, and dendrite growth.

There will be a test at the end to see if you got all of these. If you didn't the only reason I went through all of these is to let you know no one knows what ELF does, but everyone knows it does some pretty squirrelly things. For people us to just sit back and ignore ELF transmissions is foolhardy, so let me go back to Chemtrail again.

Chem-Trails in Action

Let's leave the HAARP transmitters and look at those pesky chemicals being pushed out the back of jets. The following chart [left] shows huge increases of metallic particulates after chem-trails were noted in Phoenix Arizona. The first column shows tested level and the tiny column to its right shows the "toxic limit" of Barium, copper, Manganese, Aluminum, Magnesium and other substances that could build a synthetic reflector of radio signals. To its right is a similar analysis in California [the spikes represent high level of specific metals found], and these tests have been taken around the country with almost identical issues as someone is pushing metal Oxides out of planes in patterns.

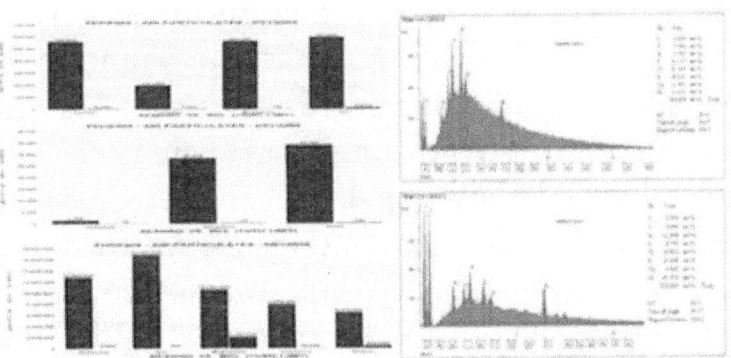

Don't consider the hundreds of tests verifying metallic components in what appear to be contrails. We got assurance from the EPA. "EPA is not aware of any deliberate actions to release chemical or biological agents into the atmosphere," the agency states, noting that the ice particles in contrails melt and evaporate as they fall to earth, posing no threat to human health. The EPA does note one potentially harmful effect of contrails: "Contrail cloudiness might contribute to human-induced climate change. Climate change may have important impacts on public health and environmental protection. That will depend on flight traffic and routes in coming decades as well as other weather changes tied to global warming." One simply wonders just how many Chemtrails are out there. I don't think anyone knows the answer to that, but there is evidence that the spraying has been going on even before the HAARP system was in place.

Solar Radiation Management

Solar radiation management [SRM] is a group of methods to reduce radiation from the sun in the environment by spraying the radiation-reflecting chemicals into the atmosphere. The Fifth Assessment Report of the Intergovernmental Panel on Climate Change [IPCC] has issued statements regarding *the importance of SRM in fending off global warming*, but most government agencies are still not coming out a stating the obvious. In the olden days, the metallic shield was thought to reduce the amount of heat that could penetrate so there could be relief from extreme heat. That seems to be a noble and worthwhile objective. Unfortunately no one told citizens what was going on just in case something bad happened so there have always been rumors. Today we see these things just about every day in some parts of the country. Here are a few examples. The group following shows how many mystery "trails" start and stop and start again just like they were being controlled.

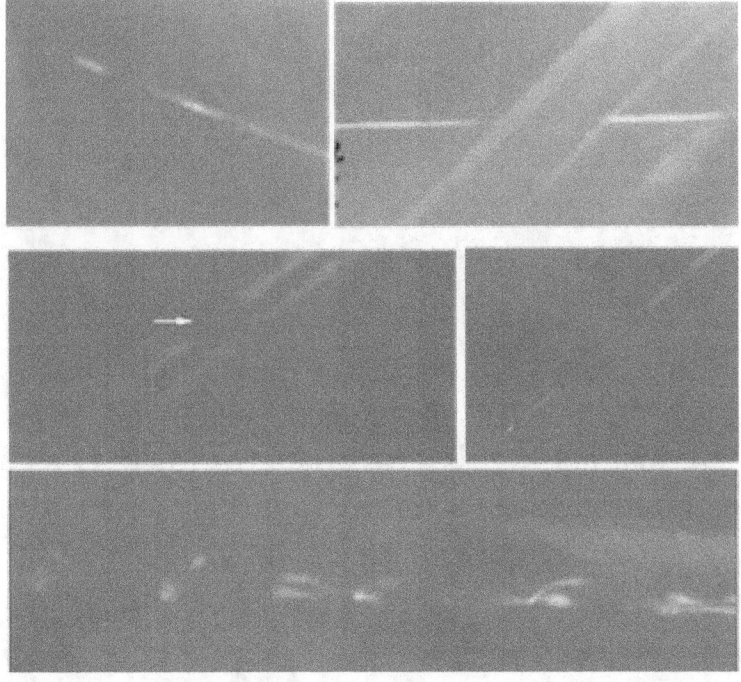

Just about everywhere you look you can find these stop-start-stop-start trails that cannot be condensation trails.

The next group are more chem-trail patterns to provide needed reflection of particular frequencies. Possibly, they are just put in crisscross patterns to cover as much sky as possible and do the thermal blanketing for solar radiation management, but there are just so many.

There is no way the Chemtrail have not been noticed and almost assuredly a great number have data concerning the issues brought out here, but almost every organization continues the ruse so they can continue to rake in money, funding, accolades, influence, and all the rest from the CO_2 that isn't doing anything but making hazy skies when it gets too concentrated. It is a crime to the entire world. That fact does not mean no one has noticed and trying to do something about it. Let me just provide what is being done privately. *"Over the last six months a tireless effort to organize solid legal action for the purpose of exposing and halting global geoengineering programs has progressed behind the scenes. The toxic heavy metal fallout from the ongoing aerosol operations has been confirmed with dozens of lab tests from the U. S. and other locations around the globe. Environmental protection agencies have not disclosed this verifiable contamination which has catastrophic implications for the environment and human health. A legal filing is being prepared to expose this fact and thus the secret and illegal global climate engineering programs that have been carried out for decades without public knowledge or consent. A team of 11 attorneys (8 from the US and 3 from Canada) are working together with the largest and most visited independent informational website on the subject in the world, GeoengineeringWatch.org".*

California Drought

I want to share one of the conspiracy theories with you as I think there is some merit in its characterization and this has to do with the high pressure area that seems to be staying off the coast of California. In May 2013, a state of drought was declared in California that persisted throughout the rest of the year, the state's driest year to date. In December, a massive wildfire broke out near Big Sur, thought to have been spurred by the dry spell. More than 900 acres of land were destroyed in the blaze. The drought continued into 2014.

Here Comes the Conspiracy

The logical explanation given by scientists was that the drought was the result of global warming, man-made climate change, or both. Conspiracy theorists are convinced that geo-engineering is to blame for the drought. This gets a little bizarre, but according to theorists, geo-engineers have cut the rainfall in California with the continuous spraying of aerosols and use of ionosphere heating. By turning California into a desert, its citizens will be at the mercy of the government to supply food they can no longer provide for themselves, leaving the government in total control of the population. Proponents of the theory are even going as far as to say that there is no natural weather anymore. They believe that continued spraying has caused the planet's natural climate system to stop functioning. Now, the geo-engineers are simply making up weather patterns as they go along, hurtling America into a state of weather warfare. One thing is for sure; for most of the winter massive crisscrossed chem-trails were seen off California's northern coast around the location shown below so here goes a modified possibility.

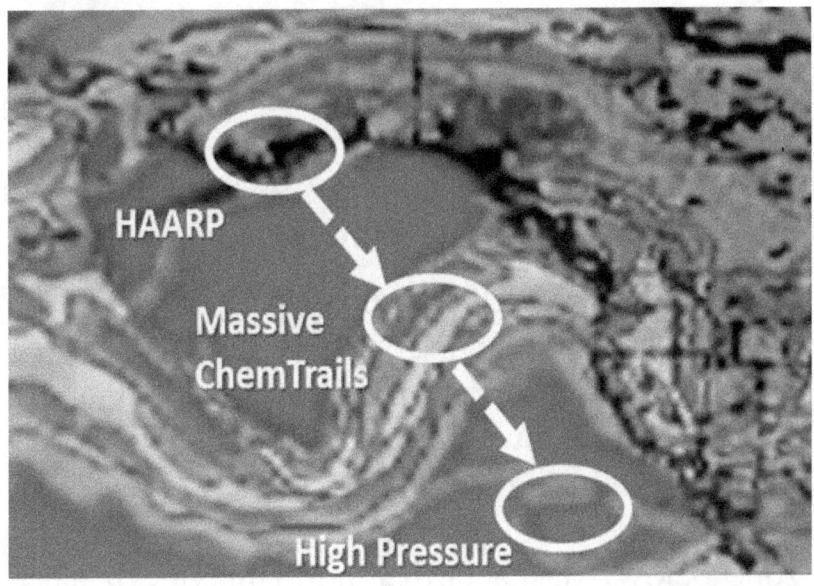

Just to sort of show you what I mean about the Chemtrails, the first satellite image shows disorganized Chemtrail images off the coast of Oregon in 2003 and the more well-known crisscross pattern off the coast of Northern California in 2012.

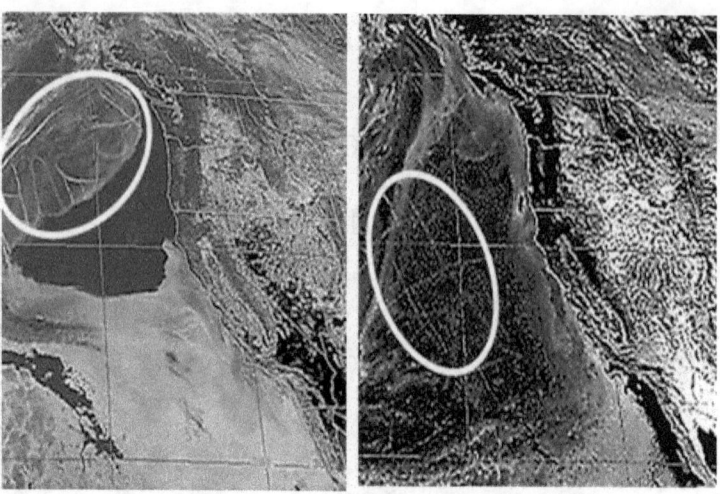

Please notice that the location of the mysterious high pressure area is about the same distance from the chemtrails as the HAARP transmitting site. So What!!!! Well let me give you a few possibilities.

- What if the HAARP transmit energy bounced off a metallic oxide reflecting surface to increase the air pressure near the bottom of California? I know it sounds absurd, but as you look at the next set of images, I think you will see that pulling the high pressure line north, generally pulls the low pressure zone below Alaska farther north making it easy to build a high pressure area in the empty void.
- If the HAARP transmit energy increased the air pressure slightly, it could also increase the pressure at its location.

That brings us to the sequential, dynamic, tracking map shown next. I put a dotted line across the main constant pressure line to make is stand out in black and white

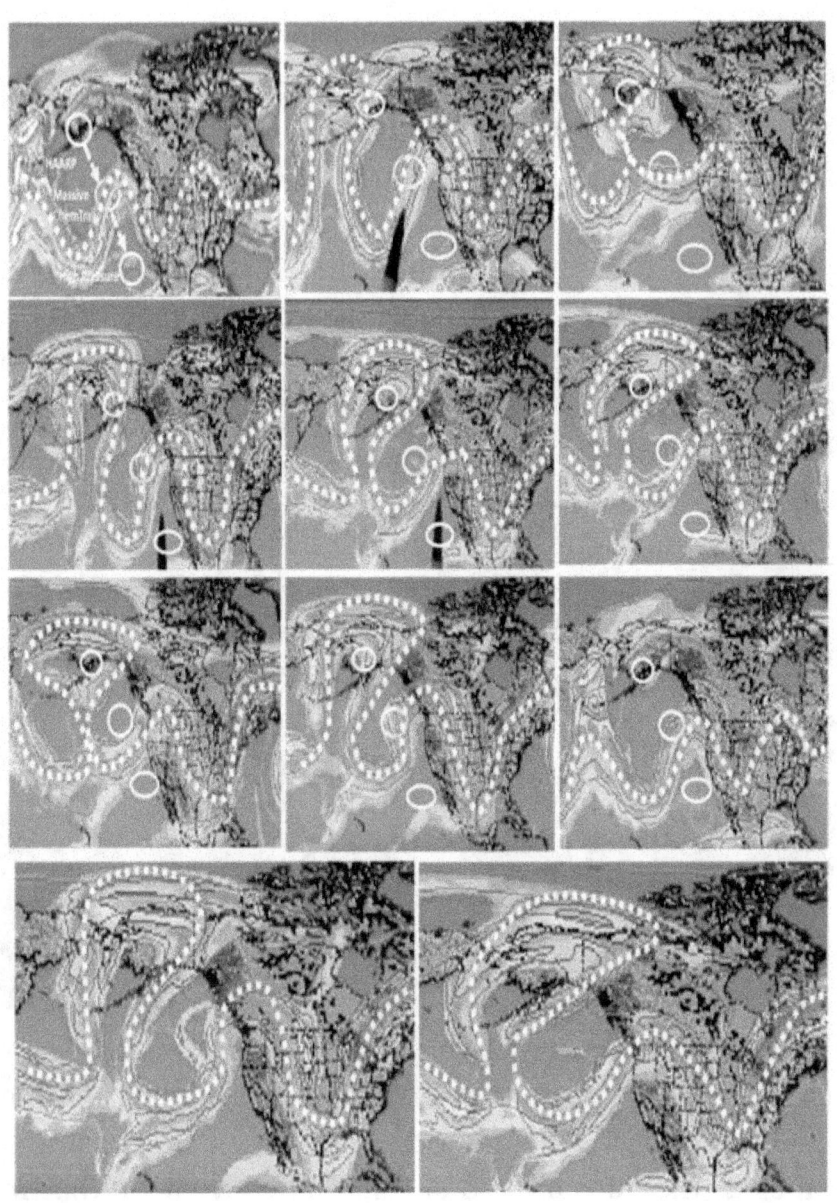

This group of images shows the winter weather pattern from about December until about March of 2014. The first image shows the normal polar jet cycle I mentioned before but images 2, 3, 4, 5, 6, 7, and 8 show the thermal gradients are grossly pushed into the Artic which warmed up the ice and helped melt the glaciers. Image 9 is back to normal only to be distorted again in images 10 and 11. There is no telling how often this happens, whether it is a purposeful activation of HAARP of if it is an unintentional reaction from an experiment. But the distortion is focused around HAARP. Speaking of distortion; what is I told you about a hole in the sky over the HAARP transmissions? Please look at the following image from a passenger jet of a hole being driven through the clouds. Does that look like something that might be dangerous?

When we look at a very small Ionosphere heater located off the coast of California, look at the strange cloud formation as high pressure zones make strange shapes as they are moved around. The arrow on the first image shows where the transmitter is off the coast of Baja California.

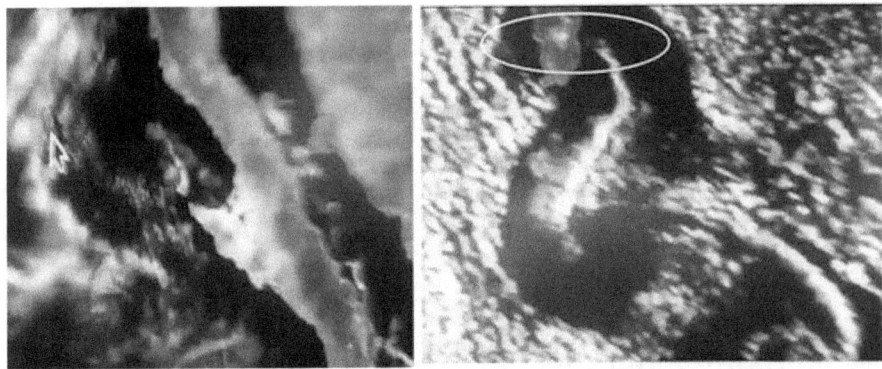

You can make up your own mind. I just want you to see the data, but another thing should be addressed as well. It is the blasting of the magnetosphere with what are called "hot Electrons by the HAARP. Whether or not the Chemtrail coincidence added to the disruption or the HAARP transmission pulling the temperature farther north on its own, no one knows, but one thing is for sure CO_2 did not cause those pressure patterns to move that way and continue to move when transmissions are initiated at the wrong pressure cycles.

Ionized Shield in the Magnetosphere

To find out about this one we need to review one of the Environmental patents of one of the HAARP staff. His patent describes a method using a radio beam at between 1 and 3.6MHz that is aimed along the magnetic field lines of our planet. [Earlier I called this modulated heating of the auroral electrojet]. It seems that the fast turn on of the RF beam shoots these "Hot electrons" into the Magnetosphere as the dangerous ELF modulation is applied which then drive the hot electrons upward into the "magnetic envelop" called the magnetosphere where they remain trapped. Get enough electrons and you can bounce signals off them like a mirror. I should also mention that the RF beam itself acts like a huge antenna below the magnetosphere, But one has to wonder what happens to our magnetic shield against cosmic and X-rays from the sun? Additionally why would you want the Magnetosphere full of these electrons unless you were doing something with the Sun as almost no one really uses the magnetosphere to talk to someone. You waste all your signal just getting up to the magnetosphere and once there, there are all types of interference elements like cosmic and X-Rays. The plot thickens.

Alfven Resonance Experiments

The earth has a natural electromagnetic resonance. In fact Nikoli Tesla used this resonance to send electricity without wires through the ground. Still an almost impossible task the idea is that at a resonance, the resistance goes to zero so there is no losses. It's not that simple, but there are experiments going on to look for resonances of the earth. The easiest one to find is called Schumann Resonance which is the time it takes for a radio-wave to go around the surface of the earth. This magic number is 7.75 Hz. If you use this critical frequency, theoretically if you can get around the earth, the signal would be returning and you are transmitting so the signal will "Build" on its own and people could "hear" data on this path all the way around the world. Let me just say this is one of the dangerous ELF frequencies, but strangely, another signal is "Heard" when this is attempted by HAARP and the others. Strangely this new signal is about 0.9 Hz and is now called the Alfven resonance. If it's going around the world it is taking 8 times as long so the electromagnetic waves are slowing down immensely. Sending RF through water slows it a little [about 25%] but this is ridiculous and no one knows what it means. We can believe there are experiments going on at these dangerous ELF frequencies today without knowledge of what could happen to our environment as we should consider that the signals are going through the earth rather than around it. Going through magma would slow down the signal, but if you excite magma is it a bad thing? Another concept is that the waves travel through the thin plasma between the top of the ionosphere and just below the magnetosphere several thousand miles above the surface. The RF has farther to go and the plasma would also slow the RF considerably, but what happens if you excite plasmas in between the Ionosphere and Magnetosphere? The reason I am bringing this stuff up is that there is experimentation going on at all of the Ionosphere heater locations and some of this experimentation would surely affect our Earth's weather patterns but still the continue so they can be the first to say they know what an Alfvin resonance is. With the real possibility of the Chemtrails and Ionosphere heaters causing the global warming, let's look at more of the "Proof" others are giving to assure us they are not. There is little doubt most scientists must see a correlation between the unusual heating and the unusual HAARP radiations, but here is more data to mull over. As climatologists KNOW what our sun is doing and they ignore it because it does not fit their AGENDA. To give you a sense of what is known, the next section addresses how our sun can and does affect us.

The Sun

Sun Cycles Match Thermal Cycles-The NOAA quasi-scientists know that the solar cycles and the temperature not only go up and down in intensity, but even the 20 year cycle is approximated by thermal shifts. No one is telling anyone about this as it would take away from the various conspiracies they are instrumental in initiating. The second chart shows the greatly enhanced CO_2 rise during this same time period which shows no correlation.

Maybe I don't have a trained eye, but my thoughts are that the sun changes our planetary temperature and as it gets hotter, we get hotter.

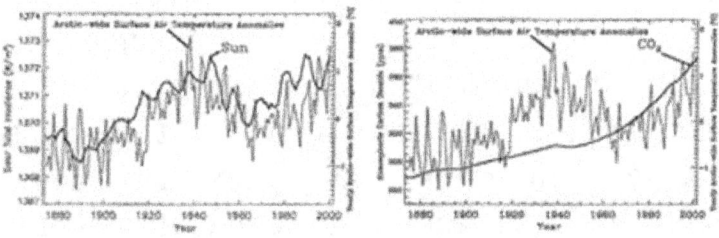

All those greenhouse gas things are the real reason for temperature fluctuations. The real culprit is the sun. All of the heat in our atmosphere is solar heat. If the sun gets brighter and burns away clouds, things happen and when the sun generates cosmic and X-Rays that hit the earth other things happen, Instead of just shining, the sun blast energy out in spurts.

Solar Flares-If we look at the correlation between temperature and total solar irradiance we see a much better relationship and we can begin to understand CO_2 is not the main player. In fact; the tiny CO_2 gas levels could not possibly affect weather no matter how many gas guzzling cars there were! It is the activity of the sun (sun spots, solar flares, modification of other galactic cosmic radiation from outer space, the effects of solar wind, and magnetic flux), that affects the radiation arriving on earth. Here is a big one. The sun moderates cloud cover! Approximately 1% of the atmosphere is greenhouse gas and 90-95% of that is water. CO_2 is about 0.05% of the atmosphere. But only 5% of that 0.05% is man-made!

Solar Brightness-Besides the magnetic field fluctuations, changes in the brightness of the Sun can influence the climate from decade to decade, but an increase in solar output by itself only changes by about 0.1% between hottest and coldest times and the current direction is just now starting to try to increase the average temperature as shown next left. The sun's irradiance has an affect on our temperature that should be recognized as shown next right. It shows that while CO2 levels are changing at a constant rate, the intensity of the sun is cyclic just like the top two Temperature lines reveal. That brings us to Cosmic Rays.

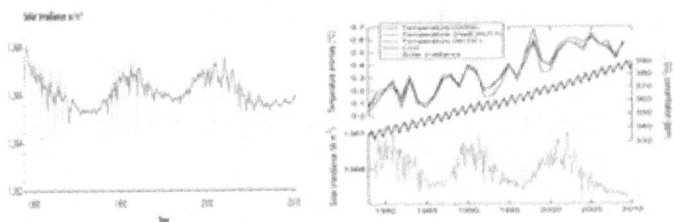

Cosmic Ray Temperature Modification-As shown next we can say global warming cycles are seem to also be associated with Cosmic Ray cycles, X-Ray cycles and Sunspots. Certainly someone would be showing this stuff to you so you would see the sun's control over your temperature, but somehow, they forgot to tell you.

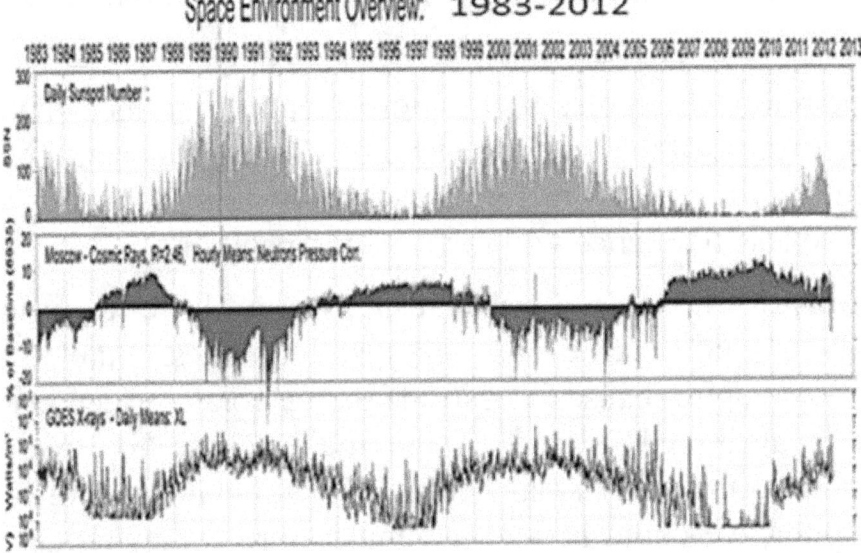

Climate Control-Attempts to correlate solar activity with global temperature have been going on for some time now as you would expect. Measurements from the SORCE's *Spectral Irradiance Monitors* show that solar UV variability produces colder winters in the US and southern Europe and warmer winters in Canada and

northern Europe during solar minima. Here is what we are finding.

Cosmic Clouds-Solar wind-mediated galactic <u>cosmic ray</u> changes, seems to affect cloud cover which would change the thermal characteristics. Much of the solar effects have been clouded by the evils of NOAA trying to keep their favorite stocks going up, but I think you can see climate is almost totally controlled outside our atmosphere. The cosmic ray change over the cycle changes clouds. Let me be direct here!

All environmental scientists know this and they also know that the most significant changes are in the Polar Regions.

Therefore, the poles show more thermal variance than other parts of the world. Get rid of Cosmic Rays and the Poles will stay more regulated, but making sure cows have less flatulence will not save the penguins

Causes of Solar Cycles-Just so you have a little background solar cycles are just like tides. Like out moon, Jupiter causes variations on the solar surface and it is attributed as one of the major cause for cycle sunspot and other solar activity. If we could get rid of Jupiter, we could make our temperature cycles reduce, but eliminating coal burning just removes someone's livelihood.

Sun Spots-The sun is 1.3 million times larger than the earth. It only makes sense that when its temperature changes, our temperature changes just like I showed in the previous charts. Well, think about it. Every year, the temperatures rise and fall with spring, summer, fall, and winter. A year is simply a 365-day cycle. Every day, the temperatures rise and fall with daytime and nighttime. We can neither change them nor stop them any more than we can stop the Earth's rotation. The temperatures fluctuate based on these cycles. So clearly, the Earth's temperatures rise and fall based on its exposure to something we call the sun. There are larger cycles of the sun called "solar cycles." The following graph left shows the cyclic nature of sunspots and this solar cycle thing. Notice that as the temperature increases, the amplitudinal difference of the solar flare cycles increases.

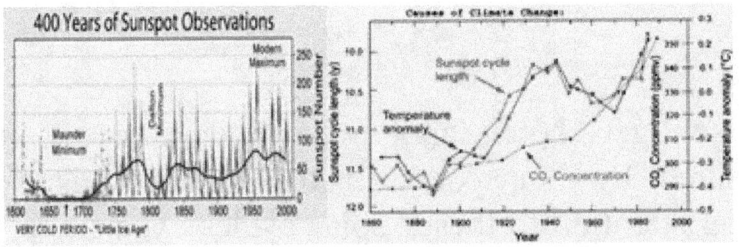

From the above graph right we can see more detail. From 1978 the temperature is getting slightly warmer in parts of Greenland, so what did we do differently before 1978. As before, the graph shows that there has been <u>no appreciable slowdown in the increase of CO_2 in the atmosphere</u>, but there was a fairly significant reduction in the temperature between 1940 and 1978.

The 400 year graph following shows that the correlation does not stop. Sunspot activity is cyclic and when there are more spots, the temperature rises. Some have

177

suggested that if we could simply shield the earth from sunspots, our global warming issues would evaporate. There is a consistent similarity between sunspot activities. While CO_2 levels simply were rising since the 1800s, sunspot cycles go along with tmeperature rise and fall. Some of the Chemtrail activity, no doubt, is to regulate the temperature during these high peaks, but please understand, messing with natural weather may have unknown concequences. We know governments know this so they try to stay quiet about the weatehr modification attempts.

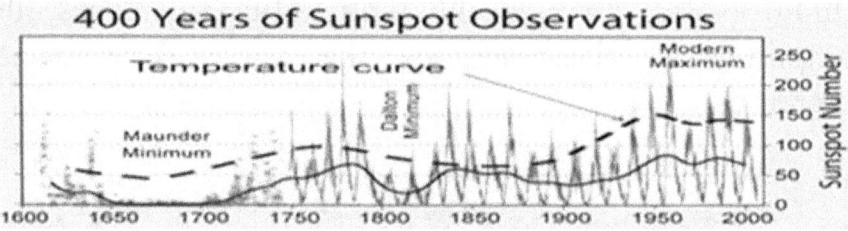

The sun gets hotter and times when it cools off as measured by "sunspots." If the climatologists were paying attention to these "solar cycles" years ago, they could have told you that the Earth would get warmer during the 1990s, and then it would begin to cool just like it has. The next graphs may help us understand some of what might be going on as the <u>magnetic field of the sun has been reducing over recent years</u> as our earth has become a little warmer.

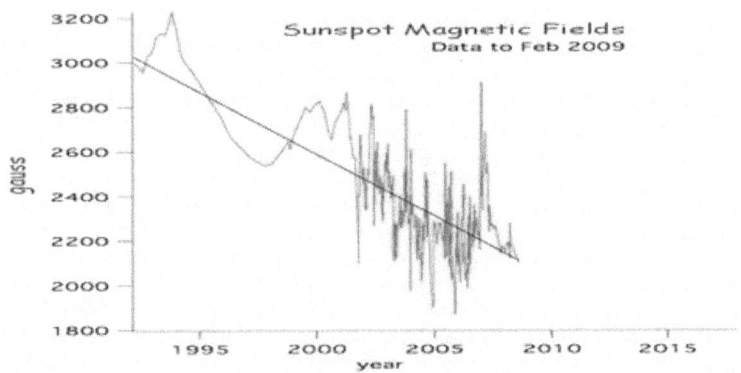

NASA's <u>S</u>ounding of the <u>A</u>tmosphere using <u>B</u>roadband <u>E</u>mission <u>R</u>adiometry [SABER] mission began routine operation in 2002. Do you sometimes get the feeling that they make up these crazy names just get a neat Acronym? This is a satellite set up to see how our atmosphere is doing. Anyway; it was found that a solar-geomagnetic storm blasted the ionosphere in 2015 and lit up sensors on the orbiting instrument in a way that researchers did not expect to see.

Corona Mass Ejection Danger

While there is little chance of a meltdown of our atmosphere in our lifetime, and Ice core timelines suggest we are head more towards the next Ice Age that is quite a few years away, there is something we had better understand right now. That is something called a Corona Mass Ejection from our sun. While alarms are everywhere for CO_2, I find little coverage over a true dilemma we are having as the sunspot cycles increase and decrease over time with occasional directed bursts. To let you know what I'm talking about we need to travel back to 1859

1859 Solar Storm Event

Some have not heard of "the Carrington Event", it was a massive solar storm in 1859 that fried telegraph machines all over Europe and North America. During this thing, northern lights were reported as far south as Cuba and Honolulu, while southern lights were seen as far north as Santiago, Chile. To make this even more scary, the flares were so powerful that *"people in the northeastern U.S. could read newspaper print just from the light of the aurora"*. That was only the beginning. The geomagnetic disturbances so strong U.S. telegraph operators reported sparks leaping from their equipment. In some areas it was so bad the sparking set fires. This was the worst recorded, but certainly not the last as these things would be extremely bad today given our new dependence on Radio waves, GPS, Telephones, Internet, and our massive electric power grids. .

- First, high-energy sunlight, mostly x-rays and ultraviolet light, ionizes Earth's upper atmosphere. This interferes with radio communications.

- Next comes a radiation storm, potentially dangerous to unprotected astronauts and others outside our fantastic magnetosphere.

- Finally would come this coronal mass ejection [CME]. It is sort of like a slow moving cloud of charged particles that can take several days to reach Earth's atmosphere. When a CME hits, the solar particles can interact with Earth's magnetic field to produce powerful electromagnetic fluctuations. Transformers would burn out, Automobiles might be destroyed and people could be without electricity of weeks or even months. Some estimates indicate loss of power for years.

2003 Event

A very light CME event happened in 2003. Called the "Halloween storms" of 2003, the event interfered with satellite communications, produced a brief power outage in Sweden, and lighted up the skies with ghostly auroras as far south as Florida and Texas. This was a tiny thing that wasn't directed at us.

2013 Event

In 2013 the sun let loose with a Carrington-magnitude CME. Luckily, the position of the earth at the time prevented the burst from hitting it as the charged particles streamed into the Solar System. The chances of additional events of such magnitude may be far greater than most people think.

February 2015 Event

SABER observed the strongest geomagnetic storm in over 10 years. It was felt when a fast-moving Coronal Mass Ejection reached Earth's magnetic field. The energy radiated by the storm, as measured by the SABER instrument, totaled 41.7 billion kilowatt hours, enough to power every home in New York City for more than 3 years. While the storm had unleashed a flare which was a million times more powerful than all of the nuclear weapons in existence combined, it was not directed at us. For what it's worth, this particular active storm region, is designated AR2192. It exploded with 2 X-class flares which caused some short-lived high-frequency radio black outs around the globe and very little panic.

2016 Upcoming Event

The storm surges of 2015 are still happening. The current storm is only 14 times larger than the Earth. An eruption on the sun at just the right time and at just the right angle could result in a society-crippling electromagnetic pulse blasting this planet. While the danger posed by a "G5 solar storm" gets mentioned occasionally at Congressional hearings, there really hasn't been any major action. . So if your computers, cell phones and electronic equipment get fried at some point, do not blame CO_2.

Why Do We Care?

After an estimated 60 years of climate engineering has decimated Earth's atmosphere and the natural protection that was once provided from it we should not think we are invulnerable. After the latest assaults into the magnetosphere and the "hot electron" bombardment, we can believe geoengineering has already left much of the planet exposed to dangerous UV radiation levels. Then there is the damage the constant spraying of particulates with chemtrails has done to the ozone layer we can assume our once robust atmosphere has taken a hit. Now, the risks from large solar flares and CME's is much more problematic. Besides all that, we must wonder if the power grids shut down for too long, are nuclear power plants around the globe that would not be able to cool down reactors.

Instead of worrying about CME's we are told to buy solar cells and all our worries about global warming will be neutralized. One problem I can see about that is that solar cells won't work well during an Ice Age.

The Next Ice Age

If the current modern global cooling continues, winters in the Northern Hemisphere and summers in the Southern Hemisphere could be colder. The signs seem to be starting already. Greenland data actually shows the temperature is turning colder as the CO2 levels are getting slightly higher as shown below.

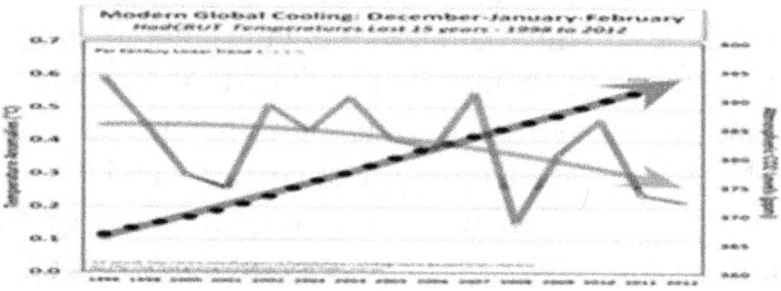

How Much CO_2 is in the Air?

Over the past 50 years, some have speculated that over 1 trillion tons of CO_2 have been thrust into the air by humans burning fossil fuels. As I mentioned before most of these Green Gases would stay in the air for thousands of years so all that trillion is in the air that we are breathing. Even if all this is true, why has there been NO substantial change in temperature cycles? It is now become obvious that the warming trend of the 80s and 90s and even into 2000 has turned around and is now beginning to cool as this massive block of CO_2 just sits there not bothering anyone.

Over the last 10 years the trend has shown a slight drop in temperature. All you hear is the opposite, but whoever cared about the truth.

Thermal Cycles-I showed the next chart earlier concerning both CO_2 and temperature change cyclic nature, but what happens when we just look at after the Pleistocene Extinction?

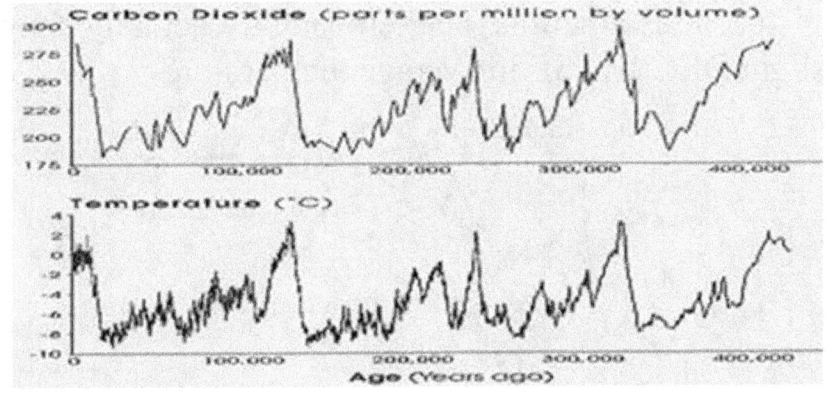

I spread the graph out from about 20 thousand years ago until the present for the Antarctica Ice cores

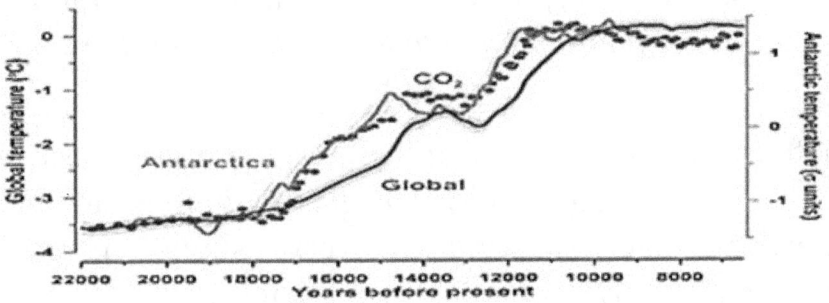

Notice 10 to 12 thousand years ago, the temperature finally stabilized and has slowly been drifting colder

where the Ice core samples were taken near a massive lake toward the middle of the continent. Notice the little dots describing CO_2 levels. Sometimes they are behind the thermal changes and just as many times they are ahead of a change. It is as if Hydrocarbons had almost no effect. When we look at ice core sampling from Greenland we find an even larger anomaly. I superimposed the Antarctica thermal graph on top of the Greenland one. [Antarctic graph is the smaller of the two.] Notice that they are very similar after the Pleistocene Extinction. The temperature actually shows a slight DECREASE in average temperature.

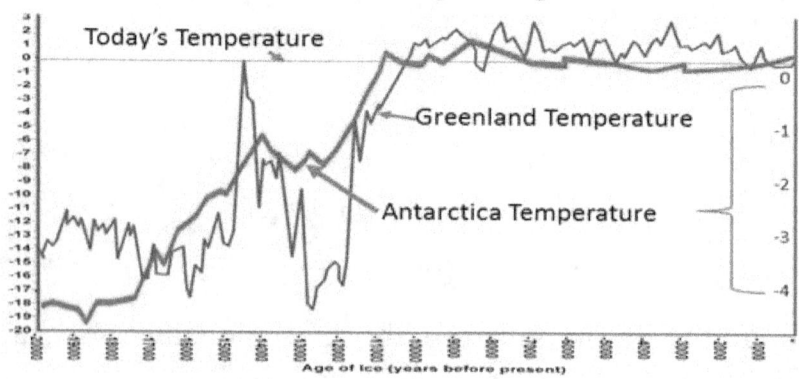

Like I stated before, position of the data is everything. We could get another sample to show a slight increase as well.

Another issue with all of this is that there are massive swings in both directions. The next graph is another Antarctica one showing a very slight increase intemperature over a 400 year period that is less filtered.

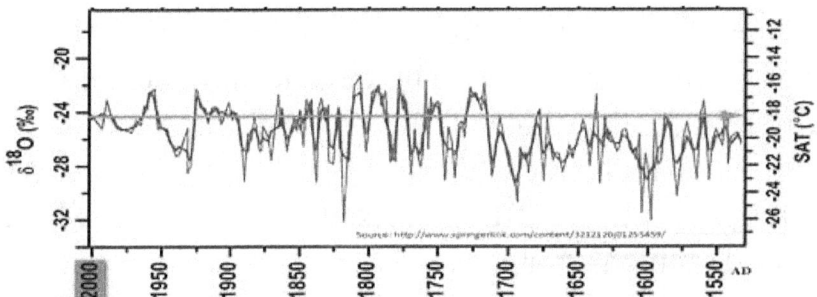

Please notice something about this graph. Nothing seems to change the course of the slope for any period Certainly it shows there were massive temperature drops during World War II as we began huge manufacturing of warproducts-----Wait a minute!!! That is the wrong direction as hydorocarbonated air didn't make the golobal warming thing. After the war, in the 1950s we see an increase as massive reduction in manuacturing was noted and so on. If global warming is not caused from hydrocarbons, what are we seeing?????? The answer might be on the sun.

Let's get real! The Earth doesn't seem to be poised to turn into a pressure cooker. Most likely, what is really happening is just as bad as global warming. We are evidently going steadily towards our next Ice Age.

Many climate experts believe we are overdue for an Ice Age so they check for indications continuously. The ice core studies and long term weather research all indicate that the Earth alternates between Ice Ages and inter-glacials. In 2002, at the South Pole, the penguins were stranded because of an abnormal ice buildup not because of the blazing heat the greenhouse model has suggested, but instead the weather was colder. This is only one of many indicators of the impending problem. The graph following left is the deuterium concentration at different depths of ice. There are two things to notice. The first is that since about 11 thousand years ago, <u>Antarctica has been getting colder.</u> We have had a base on the continent for decades and have just had to reposition it as it is now too far from any open sea to allow for reasonable use. The slope represents about an 8 degree drop in average temperature over that time period. The only significant longer term thermal events were about 6000 years ago, 10000 years ago, and 11 thousand years ago [that whole Venus incident]. This brings us to the second thing to look at. There are thermal spikes that last for 100 years or less. Although short lived, the thermal change during these times is as much as 3 degrees centigrade. That factor is significant when viewing a later chart.

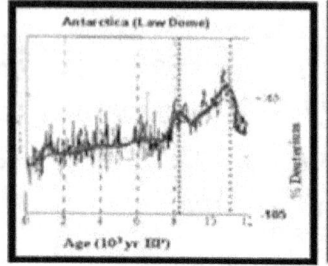 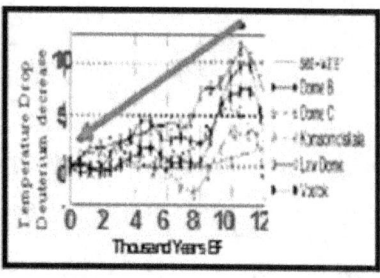

The second chart, right, shows the temperature of Antarctica over 5 different sites. By cross comparing the data, we can be assured that the data has a high degree of accuracy. Antarctica and the entire world **are** getting colder. Instead of the long term ice core graph being shown to us, the short term one below is typically provided to make us think the opposite is happening so we don't use our much needed coal reserves.

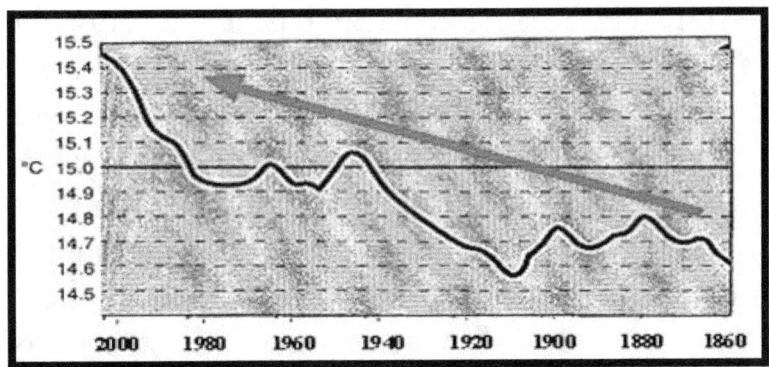

I've been complaining about the halting of all "fluorocarbon sprays" and the testing of "methane producing cattle flatulence" because of the fear of the greenhouse effect, as it is just a way to steer money to "green" industries those jumping up and down have money in. I'll bet you thought that there was "Global Warming" all over, because everyone kept pushing it into your head, but you have been lied to.

This wasn't on purpose, necessarily, but it shows the problem concerning how the science community gets so wrapped up in a theory, that they will do almost anything to show that the theory works.

Here is where there is a lot of conflicting information. Some scientists still hold on to the warming theory and papers are still being written about it every year. If they write enough papers the earth will get warmer, but they will have to write very fast to cause enough friction. Some of the data does suggest a very slight warming cycle, but more of it does not go along with this theory and <u>there is NO data that suggests that anything we do about it has any affect.</u> In all the millions of years of this cyclic behavior, the earth has **never** gone into a significant "Global Warming Fiasco". It's like saying Mars could, all of a sudden, get hot. It won't happen. Our planet has a hard enough time keeping itself warm for us. Remove the atmosphere or disrupt it in any way and <u>the temperatures will plummet</u>. You have seen information in newspapers and articles about the earth warming, but what if we are going the other way? It makes more sense and we need to wake up.

Destroying Trees

Scientists still don't know what even causes Ice Ages much less can they devise ways to slow its progress. We know that the bogus warning for people to halt cutting trees or the earth will get too warm didn't halt tree cutting. The idea of having too much CO_2 because there are not enough trees to consume it is a bad thing, but like the cattle methane, it should not be brought out as a reason for the earth getting warmer. The earth is still doing what it wants to do.

Axis of Rotation

One thing that does absolutely cause the homeostasis of the earth to change is a change in its axis of rotation. If a change like that occurs, plants won't grow as much because of the temperature changes and other environmental changes that disrupt growing cycles. Less coverage of the earth means the earth temperature will drop in temperature significantly rather than increase and large masses of ice would be repositioned from one location to another which would further amplify a delicate situation. This may be one way an Ice Age begins. I showed the hot spot indications of Hawaii and how they line up with major temperature shifts, but around the time of the shifts, there were huge temperature fluctuations as the Earth did not settle quickly in it new position. According to history, the last major shift was only 10 thousand years ago so we are possibly safe from that occurrence for the time being. We are in the region of those violent shifts "after an axis shift". Our Earth is slowly establishing variations on climate associated with the end of the Pleistocene shift. I remember form a movie when Tom Selleck asked a Tibetan Holy Man about what life was. The answer was –

The Oxen is slow but the Earth is patient!

The earth takes a long time to settle. By the way I'm all for halting the extermination of our rain forest and reducing the amount of CO_2 burning in urban area, because I like the medicines we get from the forests and I like the oxygen made by plants. Fewer trees are not good for any of us and industries should be encouraged to build away from cities. That being said Ionosphere heaters should be more cautious; Chemtrail thermal blankets should be voted on; NOAA and IPCC should both be fined when they are found to lie rather than be praised; and we should continue to use coal.

Conclusions

Hopefully you got something out of my rantings here. If you didn't get anything else out of the book, I certainly hope the following elements are being considered in a stronger way. Just because a science community has endorsed the manmade Global Warming issue, don't believe all you hear and see, even when there are warnings of prosecution for those who don't bow down to their conclusions. Many quasi-science organizations lie. While I picked on a few, most have ulterior motives that don't always include telling the truth. Especially once they take a positon. Hopefully you are thinking twice about the following "KNOWN" characteristics.

- Don't worry about the dangers of CMEs from the sun after all people do cause the sun to do stuff.

- Don't worry that ELF signals are being shot into our Magnetosphere. We have no idea what they do so why in the world is there anything to worry about.

- Don't worry that the Ionosphere heater experiments are making visible plasmas in the air. That does not mean anything and the idea that the experiments try to heat the Ionosphere shouldn't worry you either. After all it is way up there.

- Sugar sat on a rock and finally combined to make DNA. Then this DNA goo was hit by lightning and it became alive. While there is absolutely no difference between a live and a dead DNA do not for one minute believe there is a creator of life.

- Nuclear decay timing is the best way to determine how old something is because a half-life is constant.

- Dinosaurs lived millions of years ago and those found that are radio-active and not fossilized should be ignored.

- Men in the Pleistocene were little more than standing apes. The multitude of historical references of the wars should be ignored.

- The earth axis is constant as our planet is stable. Just ignore the hundreds of thousands of mammoth carcasses in the Artic Siberia.

- Venus is our major example of what using too much CO_2 would be like. Ignore how it was almost split in half and its signs of war.

- Venus suffered meltdown millions of years ago. Don't look at how the pre-meltdown details of the planet all show it was a very recent event.

- While CO_2 can't absorb solar energy and makes up 0.2% of the air, it is causing the Earth to heat up.

- Just because effluent from many jets is being placed in geometric shapes; is filled with metal oxides; stays visible while other contrails dissipate; and has been seen coming out of the back of jets that have the turbines on the wings- does not mean they are not just plain old condensation.

- Just because temperature changes increase and decrease in perfect synchronicity with solar action does not mean cars aren't causing them.

- Just because Antarctica is experiencing it coldest times ever does not mean we aren't burning up because people are burning coal for electricity.

- It is a good thing that our congress has provided billions in guaranteed funds, price offsets and incentives, grants to allow easy Green industry expansion rather than protesting our natural coal and oil resources and industries. After all, those industries are making profits without having billions pumped into them.

- Don't worry that almost 300 Coal projects have either been forced to halt or have been disapproved in favor of inefficient solar power, ecco-unfriendly wind power, and our government is forcing the elimination of Coal use even though it will cost the US a fortune to convert to much less productive energy sources.

- EMAIL traffic obtained with FOIA orders showed massive manipulation of temperature data to make it look like the temperatures were increasing. Even after the details were brought out, the temperature charts were not fixed by NOAA. Showing NOAA didn't need Satellite data, Buoy data, or historic data

194

to have a gut feel that man's civilization is killing our planet.

- Just because climatologists have made almost 100 temperature tracking and plotting models and all show the Earth should already be a dead planet, we should not quit trusting them. After all their work is keeping money in the pockets of horribly run businesses and those who pretend to run them.

- Just because the cyclic thermal cycling of the Earth shows we are heading for an Ice Age, we should still halt Coal and oil industries from being successful.

- Just because the oceans did not rise and temperature levels around the world stabilized without eliminating coal, we should still do it and we should make sure Al Gore does not lose his Nobel Peace Prize.

- While huge hydrocarbon producing nuclear wars have not caused global warming, and neither did dinosaur flatulence, we know man is now heating the earth.

- While there is absolutely no question that the NOAA and IPCC scientists know that CO2 does not leave the air for thousands of years, we should accept their addition of aerosol CO_2 to Ice Core data to show the horror of CO_2 runaway.

About the Author

Steve Preston is a long time author of scientific, esoteric facts. His series on the creation of mankind is shown below. The series focuses on the painful truths rather than whitewashed details that make us comfortable. If you are interested in the truth instead of comfort, please continue to read and, while you are at it, review other works by Mr. Preston as shown below. Like this one on global warming errors, most of the books are not politically correct and most have readers who hate the same books that other truly love. He is only interested in finding the truth rather than making people feel good about what they thought was truth for years.

Eight Part Series "History of Mankind"
The First Creation of Man
The Second Creation of Man
The Creation Of Adam And Eve
The Antediluvian War Years
Man After the Flood
A Closer Look At Ancient History
A New View Of Modern History
The 20th Century To The End Of Time

Other Works

Kingdoms Before the Flood
Egyptians In Ireland
Behind the Tower of Babel
When Giants Ruled the Earth
Ancient History of Flying Lizard People
Who Really Discovered the Americas?
America's Civil War Lie
Living On Mars, Venus, and the Moon
Disgusting Display
Vibrational Matter
Adam's First Wife
Allah' God of the Moon
Anakim Gods
Closer Look At Genesis
Walk Through Time or a Wall
Moses Saved Egypt
Biophotonics and Healing
The Devil
Mystery of Photons and Light
Mysterious Pyramids
Fast History of MILES Training

World War Zero
Why the King James Bible Failed
Anthropic Reality
Victory of the Earth
Races of Men
Our 10-Dimensional Universe
Four Armageddons
Not from Space
Creation and Death of Dinosaurs
Self-Soul Spirit
Driven Underground
God Didn't Make The Ape
The Antichrist
DNA of Our Ancestors
Strange, Powerful & Dangerous Women
History Confirmed By The Bible
Mysteries of the Exodus
Complex Earth
World War Before
Meaning of Life and Light
The Book Of Odd
Why Are There So Many Anomalies?

Thanks for Reading!

www.ingramcontent.com/pod-product-compliance
Lightning Source LLC
Chambersburg PA
CBHW070235190526
45169CB00001B/190